Introduction to the Biophysics of Activated Water

Igor.V. Smirnov
Vladimir I. Vysotskii
Alla A. Kornilova

Universal Publishers
Boca Raton, Florida
USA • 2005

Introduction to the Biophysics of Activated Water

Universal Publishers
Boca Raton, Florida • USA
2005

ISBN: 1-58112- 478-3

www.universal-publishers.com

TABLE OF CONTENT

1. INTRODUCTION TO THE BIOPHYSICS OF ACTIVATED WATER

1.1   STRUCTURE AND PROPERTIES OF A WATER MOLECULE

Water is the foundation of life on Earth. The first live organisms emerged there. The composition of the intracellular fluid of any live organism is identical to the composition of the primary ocean. Every human being spends the first nine months of his life before birth in water. Since a human body in middle age consists of water by 65% (muscles contain about 75% of water, while the blood – about 80%), it would be rightful to say that we – like our ancient evolutionary ancestors – continue to live in water past our birth. Children normally have a little bit greater quantity of water in organism, and old men have a little bit less. If the amount of water in a human body were reduced only by 10% in relation to the normal, a human being would perish.

Therefore, water accompanies us throughout the entire life, regardless of whether we habituate in the Arabian desert, the mountains of Tibet or New York's Manhattan.

At the first glance it appears that we know everything about water. In reality, water is one of the most mysterious substances.

Water's structure has been investigated for the last hundred years. The scientists, who have contributed significantly to this field include Bernal and Fowler (1933), Pople (1951), Frank and Wen (1957), Pauling (1959), Nemetry and Sheraga (1962), Davis and Litovitz (1966), Sceats and Rice (1980), Dahl and Anderson (1983).

Before analyzing specific properties of the water structure in macroscopic volumes, let's examine the structure of an H_2O molecule. Three nuclei form an isosceles triangle in an H_2O molecule with two protons (H^+ nuclei) in its foundation and a nucleus of oxygen at the top. Interaction between hydrogen and oxygen is achieved thanks to the S-P-link by two S-electrons of oxygen. The distance OH (length of the triangle's sides) in a molecule of water steam is 0.957 A. It is a little bigger in ice, equal to 0.99 A. The distance $H^+ - H^+$ in a molecule of water steam is equal 1.54 A. Ten electrons surround the three nuclei in a water molecule. Two electrons are mainly located close to an oxygen nucleus.

Several works (Lennarri-Jones, Pople, 1951, Pople, 1950) used the precise method of molecular orbits for studying the electronic structure of water molecules. In those works, it was demonstrated for the first time that conditions of the other eight electrons might be described as movements along the four oblong ellipsoid orbits. The axes of two of those orbits are oriented along the direction of the link O–H. The axes of the other two electronic orbits lie in the plane passing through an oxygen nucleus and perpendicular to the plane HOH. The axes of the four ellipsoid orbits are directed towards the tips of the tetrahedron with the center in the middle of the water molecule. The electrons move along the oblong orbits in couples (according to orientation of spins). The protons H^+, located inside the two orbits, are linked to two poles of positive electric charges in the peripheral part of the water molecule. The electrons moving along the other two orbits form the so-called lone-pair electrons, which play a very significant role in interaction between water molecules in ice and water. These electrons explain a comparatively bigger value for electronic density in the peripheral part of the molecules of water, opposite to the other part, where hydrogen is located. Two poles of negative charges of water molecules are linked with them. The currently adopted model of a water molecule is practically completely identical to the model, which had been first proposed in the work (Bernal, Fowler, 1933).

According to calculations accounting for all the peculiarities of ionic links (Coulson, 1965), it was demonstrated that the hybrid-forming process involving $2S-$, $2P_y-$ and $2P_z-$orbitals of an atom of oxygen during formation of links with oxygen atoms is a more important factor than ionicity. Such hybrid-forming process results in spatial redistribution of charge and altering the shape of electronic orbits participating in establishing of a link.

Presently, intracellular distances and angles for a water molecule are quite well known. Actual values of electric charges are somewhat less definite. A positive charge of an oxygen nucleus is almost completely screened off, while proton charges are screened off by more than 70%.

Certain parameters of a water molecule H_2O, as well as other similar molecules of intermediate (HDO) and heavy (D_2O) water are presented in *Table 1.1* (Krasnov, 1968)

Table 1.1

Molecule	H_2O	HDO	D_2O
Length of link	0.9572 *A*	0.9571 *A*	0.9575 *A*
Molecular angle	104.523°	104.529°	104.474°

Some authors provided different data for these parameters from those presented in *Table 1.1* (e.g., (Zatsepina, 1998)).

Each water molecule has four poles of electric charges (two positive and two negative), which form a tetrahedron. In order for such model to have a dipole moment of 1.87 Debye (such moment corresponds to experiments on water polarization), charges concentrated at the ends of the tetrahedron should be equal to +0.171 *e* (*e* – an electron's charge).

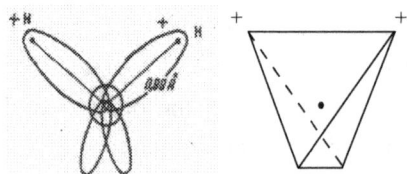

Fig. 1.1 Model of a water molecule

An adequate description of a water molecule can only be done by quantum mechanics. In the general case, a precise solution of the corresponding three-dimensional Schredinger equation presents a very formidable task. There is one possibility, however, which makes that calculation not only simpler, but more graphic as well. That possibility is based on the assumption that the so-called adiabatic approximation could be used here, which would be possible if different degrees of freedom need a widely fluctuating specter of energies required for their inducement. In the case of a water molecule, an electronic subsystem and an ionic subsystem could be identified and, ultimately, movement of the whole molecule could be studied from the perspective of a separate object. These three subsystems determine the electronic, fluctuating and rotating degrees of freedom of a molecule. Characteristic energies of inducement for these degrees of freedom correspond to the ultraviolet (UV), close infrared (IR) and remote infrared (as well as microwave) frequencies of the electromagnetic specter.

Big difference of frequencies between each of these subsystems allows neglecting mutual interference in their interaction in calculations. Therefore, electronic transfers determining the main maximum of absorption of water molecules in the ultraviolet range of the specter have no significant effect on conditions of fluctuations of a water molecule, which parameters are represented by the sum of harmonic oscillations and characterized by frequencies, belonging to the close infrared area. Accordingly, vibrating conditions have a weak effect on the rotating specter, which frequencies lay in the far infrared range and the range of sub millimeter frequencies. Several important characteristics of a molecule such as the inertia moment, the dipole moment, mass, atomic link force can be extrapolated from the analysis of the rotating specter.

The centrifugal force, generated by rotation of a water molecule, causes its deformation. For example, when rotation is accelerated to the state corresponding to the quantum number $j = 11$ (in this condition, the wave number of radiation of a rotating molecule $k = 1/\lambda = 280 \ sm^{-1}$) the molecular angle between directions of links between the nucleus of oxygen and the protons (HOH) decreases from the initial value of $104°27'$ to $98°52'$ while the distance of a link OH increases to $0.964 \ A$.

Deformation of a molecule of water may also be attributed to agitation of its electronic levels, which alters distribution of the electrons and, subsequently, to different constants of links. For example, during irradiation of water steam molecules by various specter lines of vacuum ultraviolet light with wave length $\lambda = 1219 \ A$, the length of the OH link increased to $0.067 \ A$ and the molecular angle increased by $8.3°$. Accordingly, irradiation with a different wave length (at $\lambda = 1240 \ A$) caused the angle to increase by $5.5°$ (Bell, 1965).

Condition of a fluctuating specter also affects the size of a water molecule – it increases with more agitation. This is due to an obvious fact that with increasing amplitude of fluctuations, non-symmetric inharmonic oscillator properties begin to manifest themselves. The non-symmetric inharmonic oscillator is created by ions of equal mass, which always leads to shifting of ions from their middle position.

These factors show that even without considering the specific nature of water molecules' property to form a rigid quasi-crystalline macro system (consideration of that aspect is provided below) external non-ionizing electromagnetic fields significantly affect the parameters of water molecules. Taking this into consideration explains some features of the effect that non-ionizing electromagnetic fields (for example fields of radiation emitted by mobile telephones or fields produced by various home appliances such as microwave ovens) have on live organisms.

1.2   THE AREA OF STABLE EXISTENCE OF WATER IN LIQUID STATE

It is commonly known, that water may exist in solid (ice), liquid and gaseous state. Some of the main properties of water are presented in *Table 1.2*.

In solid or liquid state, each molecule of water is surrounded by four neighboring molecules. In the solid state molecules are arranged in such a way that they are connected by opposite polarities. In this structure, each molecule is encircled by four nearest molecules (*Fig. 1.2*).

Interaction between molecules is determined by hydrogen links. Four hydrogen links of a water molecule are approximately aimed at the apexes of a right tetrahedron. Properties and orientation of these links is strongly dependent on temperature T and pressure P.

Depending on the ratio of P to T, the existence of both overcooled water and ice with a well-formed crystalloid structure is possible. The record temperature for overcooled liquid water is −92°C may be achieved at the pressure of about $P = 2000$ *Psa*. The reason for water existing in the liquid state in such a low temperature is significant deformation of hydrogen links caused by big pressure and its higher density, which impedes its crystallization.

Table 1.2 Main physical and chemical parameters of light and heavy water
(Physical Constants, 1991).

	H_2O	D_2O
Density: Ice at 0°C Water at 20°C	0.917 *g/sm³* 0.998 *g/sm³*	1.017 *g/sm³* 1.105 *g/sm³*
Viscosity at 20°C	1.005 *cp*	1.251 *cp*
Melting temperature	273.16 *K*	276.97 *K*
Boiling temperature	373.16 *K*	374.59 *K*
Specific heat (C_p): Ice at 0°C Water at 0°C Steam at 0°C	2.038 *J/g °K* 4.186 *J/g °K* 1.905 *J/g °K*	2.202 *J/g °K* 4.23 *J/g °K* 1.68 *J/g °K*
Permittivity: Ice at −10°C Water at 25°C	95 78.5	92 78.2
Thermal conductivity: Ice at 0°C Water at 0°C Water at 100°C Steam at 100°C	0.235 *W/°K*m* 0.561 *W/°K*m* 0.679 *W/°K*m* 0.025 *W/°K*m*	 0.560 *W/°K*m* 0.644 *W/°K*m* −
Time of dielectric relaxation: Ice at −10°C Water at 25°C	$6*10^{-5}$ *s* $9.2*10^{-12}$ *s*	$9.1*10^{-5}$ *s* $11.9*10^{-12}$ *s*
Molecular magnetic sensitivity at 20°C	$-1.297*10^{-5}$	$-1.295*10^{-5}$
Refraction parameter at 20°C	1.333	1.328

There are 10 known modifications of ice crystalloid structure with hexagonal, tetragonal, cubic, trigonal, monoclinic and tetraclinic syngonies. Moreover, there is also amorphous ice.

Density of ice may be both below the density of water (with tetragonal and cubic lattice for the so-called ice 1 [ordinary ice] it is equal 0.94 *g/sm³*, staying mostly constant with temperature change) as well as significantly above it. For example, density of the so-called ice 8 with the cubic lattice is 1.5 *g/sm³* at the temperature of −50°C.

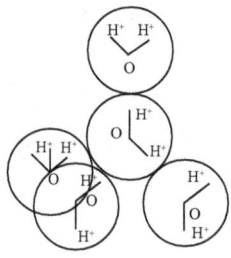

Fig. 1.2 Close vicinity of a water molecule

Higher water density during melting of ordinary ice is due to reduction of its volume caused by deformation of hydrogen links and shifts in location of molecules breaking the ideal tetrahedron configuration. Simultaneously, there is another tendency – increase of the length of hydrogen links with increasing temperature. Convergence of these two factors ultimately leads to a peculiar phenomenon of achieving the maximum density of water not at the temperature of melting of ice, but rather at 4°C. At a higher temperature, the spatial framework, characteristic of ice 1, begins to disintegrate, while "irregularly located" molecules of water begin to appear between the joints of the lattice. In the gaseous state hydrogen links disappear completely and steam behavior begins to be regulated according to the principles of gaseous state.

1.3   SPECIAL FEATURES OF INTERACTION OF IONIZING AND NON-IONIZING RADIATION WITH WATER. DIELECTRIC AND DISPERSING PROPERTIES OF WATER

It will be shown in greater detail in subsequent chapters that dielectric properties of water play a decisive role in the processes of interaction of ionizing and non-ionizing radiation with biological objects. In particular, even such a unique and paradoxical phenomenon as positive influence of small doses of ionizing radiation on live objects (called *hormesis*) can, to a great extend, be explained by dielectric properties of water and their variation under the influence of radiation. Let's consider the factors affecting the dielectric permittivity of water.

It is known, that each of the water molecules has its dipole moment p. It exists thanks to a spatial dispersion of an electric charge (i.e. the specific type of localization of the wave function of the electrons).

In the area of low frequencies, this type of dielectric permittivity of water is described by the formulae

$$\varepsilon_{WP}(\omega) = [1 + (8\pi/3)n_W \, \alpha(\omega)] \, / \, [1 - (4\pi/3)n_W \, \alpha(\omega)] \qquad (1.1)$$

In this range of frequencies permittivity is determined exclusively by the mechanism of orientating polarizing ability of H_2O molecules. The simple form of polarizing ability $\alpha(\omega)$ is determined by the Debye relationship

$$\alpha(\omega) = \alpha_0/(1 - i\omega\tau_r),$$

where $\alpha_0 = p^2/3k_BT$ – static orientating polarizing ability, p – dipolar electric moment of a H_2O molecule, $\tau_r \approx 4\pi\eta_0 <R>^3/k_BT$ – time of relaxation of the orientating polarizing determined by the viscosity coefficient η_0, temperature T and the average radius of a molecule, n_W – volume concentration of water molecules.

In particular, relaxation τ_r in clear water at $<R> \approx 1.5$ A and a normal viscosity coefficient for distilled water $\eta_0 \approx 10^{-2}$ poise is equal to 10^{-11} s.

This mechanism of orientating polarizing polarization determines the form of $\varepsilon_W(\omega)$ for water from the lowest frequencies to the frequency $\omega \approx 1/\tau_r \approx 10^{11}$ s^{-1}. Therefore, interaction of low frequency electromagnetic waves (up to waves of the centimeter frequency range) with water is completely regulated by processes of reorientation of water molecules under the influence of the wave's electric field vector and also depends on its temperature and viscosity.

The next (by the order of increasing frequencies of the electromagnetic field) mechanism of forming a particle structure $\varepsilon_W(\omega)$ is ionic polarizing ability, which maximum matches the infrared frequency range. This polarizing mechanism is connected to various molecular fluctuations of water and can be described, with a high degree of precision, by an expression in the form of the sum of four of the dispersion components (Vysotskii, 1997)

$$\varepsilon_{WIR}(\omega) = \sum_{K=1}^{4} \Delta\varepsilon_K\omega_K^2/(\omega_K^2 - \omega^2 - i\omega\Gamma_K), \qquad (1.3)$$

each one of which is determined by the main lines of the fluctuating specter with wave lengths λ_K, central frequencies ω_K and amplitudes $\Delta\varepsilon_K$ (see *Table 1.3*).

Table 1.3 Main resonance of water in the infrared range of the specter (Yukhnevich, 1973)

Average resonance wave length λ_K (*mkm*)	Resonance frequency ω_K (10^{14} s^{-1})	$\Delta\varepsilon_K$	Nature of resonance
$\lambda_1 \approx 52$	$\omega_1 \approx 0.36$	0.8	Inhibited translation of a molecule of H_2O
$\lambda_2 \approx 14.2$	$\omega_2 \approx 1.27$	0.9	Libration mode
$\lambda_3 \approx 6.07$	$\omega_3 \approx 2.97$	0.2	Deformation mode
$\lambda_4 \approx 2.86$	$\omega_4 \approx 6.3$	0.2	Valency vibrations of the OH groups of water molecules

At even higher frequencies the following and the principal maximum of interaction of the electromagnetic radiation with molecules of water H_2O is related to electronic polarizing ability and is characterized by the array of closely located spectral lines with average frequency

$\omega_{UV} \approx 2.2 * 10^{16}\ s^{-1}$, average wave length $\lambda_{UV} \approx 0.09\ mkm$ and the sum of amplitudes $\Delta\varepsilon_{UV} \approx 0.75$. The expression for the dielectric permittivity in this frequency range has the form

$$\varepsilon_{WUV}(\omega) \approx \Delta\varepsilon_{UV}\ \omega_{UV}^2/(\omega_{UV}^2 - \omega^2 - i\omega\Gamma_{UV}) \qquad (1.4)$$

The main features of interaction of radiation with water within the boundaries of this frequency range refer to processes, related to ionization of atoms of hydrogen H and oxygen O.

The spectral width of each resonance Γ_I is small in comparison with corresponding resonance frequencies ω_I. It has an intricate dependency on temperature and impurities in water. It is worth noting, that the expression for dielectric permittivity as a function of a pseudo frequency $\omega = i\xi$ is important in the analysis of the question of the influence of water characteristics on the nature of interaction between biological macromolecules and their fragments (this fundamental problem will be addressed later). This circumstance is the result of the causation principle and it is determined by analytical properties of dielectric permittivity as a complex function of frequency. This result will be examined in greater detail in the following chapters. With such a substitution, any expression with the same form as in (1.4) can be worked into form

$$\varepsilon_{WUV}(i\xi) \approx \Delta\varepsilon_{UV}\ \omega_{UV}^2\ /\ (\omega_{UV}^2 + \xi^2 + \xi\Gamma_{UV})$$

Proceeding from the obvious condition that in the vicinity of any electromagnetic resonance its spectral width is always much less than resonance frequency, which means that $\Gamma_I << \omega_I$, $\Gamma_I << |\xi|$, we finally come to

$$\varepsilon_{WUV}(i\xi) \approx \Delta\varepsilon_{UV}\ \omega_{UV}^2/(\omega_{UV}^2 + \xi^2)$$

We can see from the obtained formulae that specific expression for the width of each resonance Γ_I is insignificant and it doesn't affect the final solution.

At an even higher frequency (in the range of soft and hard Roentgen radiation with energy exceeding ionization potentials of the internal electronic cores of hydrogen and oxygen) all electrons in water can be treated as free.

In this case, the dielectric permittivity of water has the form

$$\varepsilon_{WK}(\omega) \approx 1 - \omega_p^2\ /\ \omega^2,$$

which is typical for plasma-like environments (the "plasmatic" approximation in the theory of dispersion of dielectric permittivity).

Here, $\omega_p = (4\pi n_e e^2/m)^{1/2}$ – plasma frequency, n_e – total concentration of all electrons in water, m – mass of an electron.

Beside the mentioned three main mechanisms of forming the structure of dielectric permittivity of water $\varepsilon_W(\omega)$, there are two more mechanisms, one of which is related to the presence of charged free radicals (electrons e^-, protons H^+, ions H_2O^+, H_2O^-) in water and the other one is determined by the effect of atoms H_2O^*, disturbed by thermal movement and external radiation.

There are different direct causes of appearance of the ions.

Specifically, there is always a probability of the balanced fluctuation dissociation

$$H_2O \Leftrightarrow H^+ + OH^-,$$

with thermodynamic probability determined from the Bolzman expression

$$W \approx exp(-E_\alpha/kT) \tag{1.6}$$

being dependent on temperature T and the energy of dissociation E_α. In particular, at $T = 300\ K$ the value $W \approx 10^{-10}$, indicating the presence of $n^+ \approx Wn_W \approx 10^{13}$ pairs of ions H^+ and OH^- in a unit of volume of water. This mechanism, involving the process of thermally stimulated water hydrolysis, will be investigated more closely in section 1.4.

Another mechanism is still more important, distinguished by generation of additional ions in the course of radiolysis during water irradiation with ultraviolet or Roentgen quants. All products generated by water radiolysis (including charged free radicals, disturbed molecules H_2O^*, neutral free radicals H, OH, HO_2 and molecules H_2 and H_2O_2) play a certain role in radiation and chemical transformations occurring in water solutions. Let's investigate this process more attentively (Amiragova, 1964)

With absorption of each $100\ eV$ in the course of electrolysis there is an average of about four splits of water molecules. Under the influence of radiation on water, there are, initially, ionized and agitated water molecules H_2O^+ and H_2O^*. Initial radiolysis processes, leading to generation of free radicals, are executed according to the patterns

$$H_2O + radiation \rightarrow H_2O^+ + e^-$$
$$H_2O + radiation \rightarrow H_2O^*$$

The average energy of generation of one pair of ions from a single molecule is equal $<E> = 34\ eV$. Model estimates and the results of experiments testify that with absorption of 100 eV of energy from ionizing radiation an average of 2.6 ions OH^- and 2.6 ions e^- are generated. Moreover, in the process of interaction, with absorption of the same $100\ eV$ of energy about 12 agitated molecules of water are also generated. Out of those 12 molecules about three are split by ionization and about nine – by agitation. In this case, 62% of the absorbed energy will be spent on dissociation.

Radicals generated in the course of radioactive radiation (ions OH^-, electrons and agitated molecules H_2O^*) move in the water environment by means of diffusion. Further evolution of these radicals corresponds to a chain of chemical transformations.

There are two hypothesis regarding subsequent transformations of ion H_2O^+. According to one of them, ionized water molecules dissociate with generation of ion H^+ and a radical OH, while the electron, having been slowed down as a result of interaction with water, is attached to a water molecule with generation of a negatively charged ion H_2O^-. During dissociation, that ion disintegrates into an ion OH^- and a neutral radical H.

Theoretically, another variant is also possible, when the ion H_2O^+, generated during dissociation of a water molecule, quickly captures the electron in the liquid environment, which leads to generation of an agitated molecule of water H_2O^* with high energy of agitation. Later, that molecule disintegrates into two neutral radicals H and OH.

Corresponding reactions have the form

$$H_2O^+ \rightarrow H^+ + OH$$
$$H_2O^+ + e^- \rightarrow H_2O^-$$
$$H_2O^- \rightarrow OH^- + H$$

Then, in the chain of consecutive transformations, atomic hydrogen is generated

$$H_2O^+ + e^- \rightarrow H_2O^*$$
$$H_2O^* \rightarrow H + OH$$

Currently, it is impossible to say, which of the variants has bigger probability. The two demonstrated methods of radicals' generation from ionized water molecules are different by mutual position of the radicals H and OH in space.

In the first case, the radicals OH are generated in spots where ions appear, i.e. directly along the track of an ionizing particle while radicals H are generated at some distance, equal to a free run of an electron until its collision with a water molecule. That distance is approximately equal to 150 A, which is about the length of 70 water molecules placed side by side.

In the second case, the radicals H and OH, generated during dissociation of an agitated molecule of water H_2O^*, appear close to each other.

Behavior of agitated water molecules H_2O^* is quite interesting. Absolute majority of them is agitated to levels of energy, representing the lower states with energy 5.6 and 7.5 eV. Such molecule can dissociate. In order to achieve that effect, dissociation energy of 5.2 eV is required. Radicals generated during dissociation have a small surplus of kinetic energy (from 0.4 to 2.3 eV). In the process of collision with neutral molecules of water the radicals are quickly slowed down along the length of the track

$$l \approx 1/\sigma_c \, n_W \approx 20\text{-}30 \; A$$

and cannot diverge away from each other by a large distance. Here, $\sigma_c \approx 10^{-16} \; sm^2$ – cross section of elastic diffusion of the radicals on water molecules.

Due to such a rapid braking, these radicals are capable of repeated interaction (forming combinations) after stopping. That effect, localized in space, limits the mutual location of the radicals in water and is called "Frank-Rabinivitch cells". It plays an important role in the radiolysis process. Due to that effect the free radicals, generated during dissociation of the initially agitated molecules, do not participate in later water radiolysis.

Further transformation of free radicals H, OH and HO_2 takes place in the course of several alternative reactions. In particular, when water is depleted of free oxygen, a reaction of synthesis of a very powerful oxidizer (hydrogen peroxide) takes place

$$OH + OH \rightarrow H_2O_2,$$

as well as a reaction of water reduction

$$H + OH \rightarrow H_2O$$

When there is a lot of dissolved oxygen in water, another reaction occurs

$$H + O_2 \rightarrow HO_2$$

Beside that, there are also secondary reactions in the water environment

$$OH + H_2 \rightarrow H_2O + H$$
$$OH + H_2O_2 \rightarrow H_2O + HO_2$$
$$H + H_2O_2 \rightarrow H_2O + OH$$
$$OH + HO_2 \rightarrow H_2O + O_2$$
$$HO_2 + H_2O_2 \rightarrow H_2O + O_2 + OH$$
$$HO_2 + HO_2 \rightarrow H_2O_2 + O_2$$

All free radicals and molecular products, generated during water radiolysis (ions H_2O^+, H_2O^-, agitated molecules H_2O^*, free radicals H, OH, HO_2 and molecular products H_2 and H_2O_2) play their parts in radioactive and chemical transformations in water.

Specifically, radioactive and chemical transformation processes in dissolved water solutions are explained by the presence of free radicals H, OH, HO_2 as well as hydrogen peroxide H_2O_2. Probability of reactions between radicals decreases rapidly as distances between them increase in the result of diffusion. Based on that, a clear difference is established between radicals contributing to generation of molecular products H_2 and H_2O_2 and radicals, which have avoided transformation into molecular products in the result of diffusion or a reaction with a dissolved substance. According to this model, the final yield of products of radiolysis should not change with adding more dissolved substance.

Counter to that, the yield of molecular products in the same process may alter considerably. Secondary reactions of radicals with molecular products occur primarily in clear water and they don't have a significant influence on radiolysis of water solutions.

The radical OH reacts mostly as an oxidizer and because of that many radioactive and chemical transformations of oxidization in water solutions are explained by its participation. In oxidization reactions it can act as an acceptor of electrons or acceptor of atoms of hydrogen. Oxidizing capacity of OH is determined by the oxidization-reduction potential of a system and depends on the environment's acidity. In particular, oxidizing ability of the radicals OH decreases with higher *pH*.

In most cases, a neutral radical H enters into chemical reactions as a reducer. The reducing capacity of atoms of H usually increases with higher *pH*. In acid liquors, the radical H can sometimes act as an oxidizer.

Depending on pH and the oxidization-reduction potential of the environment, radical HO_2 can act both as an oxidizer and reducer. Its oxidization capacity decreases with higher *pH*, while its reduction capacity increases.

The provided brief analysis of the mechanism of radiolysis of water shows a wide variety of chemical reactions and final products, among which free radicals are dominant. Another very important conclusion is that the complicated chain of consecutive chemical transformations shown earlier may be initiated not only by an act of initial radiological impact on a water molecule but also at the presence of certain chemical and thermodynamic factors characteristic of initial dissociation.

We shall demonstrate in the coming chapters that the influence of these factors and the possibility of reducing the effect of free radicals generated by a non-radioactive way is inseparably connected with one of the most probable scenarios of the phenomenon of positive effect of small doses (called *hormesis*).

Let's now turn to considering the effect of electrons, generation of which represents one of the stages of the process of water radiolysis. It appears at first, that the role of free electrons is incommensurably weaker than the direct effect of very aggressive heavy radicals. In reality, it is quite different.

It follows from the expression (1.5) that the contribution of "plasmatic" approximation to the overall dielectric permittivity is inversely proportional to the mass of a charged object *m*. Due to a smallness of mass of an electron m_e (compared with masses of much heavier ions of hydrogen and oxygen) it is most interesting from the point of view of its influence on the structure $\varepsilon_w(\omega)$. Obviously, the "plasmatic" approximation takes place in the area of extremely high frequencies (i.e. in the Roentgen frequency range), where electrons tied in atoms and molecules can be regarded as free, as well as in the case when electrons are actually free (in the result of spontaneous ionization or radiolysis). Electrons, generated in the process of radiolysis, after their rapid slowing down in the water, are transferred into a quasi stationary condition of a "hydrated electron" with a longer life span $\tau_e \approx 7.2 * 10^{-4}$ *s*. The life span τ_e corresponds to the process of repeated combination (deionization) of an "almost free" hydrated electron. Meanwhile, an additional contribution to the general dielectric permittivity of water made by a system of such elements may be calculated by analogy with the case of plasma-like environments.

The equation for the movement of such electrons in the presence of an external electric field $\vec{E} = \vec{E}_0 exp(-i\omega t)$, which may bear the mark of an external influence as well as have purely fluctuational features and be regulated by the features of quantum electrodynamics, has the form

$$m \; d^2\vec{r}/dt^2 + (m/\tau_e) \; d\vec{r}/dt = e \, \vec{E}_0 exp(-i\omega t) \tag{1.7}$$

After completion of transformational processes, forced movement of each of these electrons is described by the solution

$$\vec{r} = - [e\,\vec{E}_0/m(\omega^2 + i\,\omega/\tau_e)]\,exp(-i\omega t) \qquad (1.8)$$

The dipole moment created by such an electron is $\vec{p}_g = e\vec{r}$ equal and the induction vector $\vec{D} = n_g\vec{p}_g$. Here, n_g – concentration of hydrated electrons.

Using the connection $\vec{D} = \vec{E} + 4\pi\vec{P} \equiv \Delta\varepsilon_{Wg}\vec{E}$ we finally find the expression for the contribution of hydrated electrons

$$\Delta\varepsilon_{Wg}(\omega) \approx - 4\pi n_g\,e^2/m(\omega^2 + i\omega\Gamma);\ \Gamma = 1/\tau_e \qquad (1.9)$$

to the general dielectric permittivity of water.

The value of $\Delta\varepsilon_{Wg}(\omega)$ determining the influence of hydrated electrons has a negative actual part

$$Re\{\Delta\varepsilon_{Wg}(\omega)\} \approx -4\pi n_g\,e^2/m(\omega^2 + \Gamma^2) \approx -4\pi n_g e^2/m\omega^2\ \text{ with } \omega \geq \Gamma \qquad (1.10)$$

which is completely analogous to plasma-like environments.

In this case there is a minimal frequency ω_{min}, determined from the condition $\varepsilon_W(\omega_{min}) = 0$. Below this frequency (with $\omega < \omega_{min}$) we have $\varepsilon_W(\omega < \omega_{min}) < 0$.

In the range of frequencies $\omega < \omega_{min}$ existence of electromagnetic waves is impossible, since for them, the wave number

$$k = -(\omega/c)\{\varepsilon_W(\omega)\}^{1/2}$$

becomes purely imaginary. Such waves cannot spread in the environment. This "cut-off frequency" ω_{min} plays an important role in the analysis of interaction of biological macromolecules in water-containing environments (including volume of a cell). Another mechanism exerting additional influence on dielectric permittivity of water is related to the presence of agitated atoms in the structure $\varepsilon_W(\omega)$. This mechanism provides for the existence of additional resonance of electronic absorption into a softer range of frequencies than that of non-agitated atoms as well as the need to account for dispersion characteristics of a system of agitated atomic electrons in a frequency range exceeding the ionizing potential of water molecules. The number of agitated molecules of water during irradiation and radiolysis can be estimated with reference to a fact, that with absorption of each 100 eV of energy from external radiation, an average of 9 agitated molecules H_2O are generated. Durations of those agitated states lies within the range from $\tau_1 \approx 0.1\ s$ (for metastable levels with quadruple transfers of the $E2$ type) to $\tau_2 \approx 10^{-7}\ s$ (for permissible dipole transfers of the $E1$ type), given that because of the effect of "radiation capture" related to resonance overabsorption of quants in the volume of the same environment, the actual duration may be significantly increased.

The presence of hydrogen links in water causes one more specific phenomenon determining dielectric parameters of water and special features of behavior of biological objects in water. In some cases, water displays the effect of non-local polarizing. Then, instead of an ordinary linear and localized equation describing the relation of induction $\vec{D}(\vec{r})$ and tension $\vec{E}(\vec{r})$ in an electric field

$$\vec{D}(\vec{r}) = \varepsilon \vec{E}(\vec{r})$$

there exists an equation for non-local relation

$$\vec{D}(\vec{r}) = \int_v \varepsilon (\vec{r},\vec{r}')\vec{E}(\vec{r})dV'$$

This kind of relation is natural only for crystals and corroborates the fact that a component of an electric field in one part of a crystal determines induction in another part. In crystals, such non-local character is a natural consequence of the periodicity of lattice and full correlation in positioning of its elements. In water, due to a large number of hydrogen links, there is also a strong correlation of polarizing in points that are spread out in space (in a much lower volume of space, of course).

1.4   SPATIAL STRUCTURE WATER MODELS

Before we conduct an analysis of main models characterizing the structure of mutual situation of water molecules in space, let's turn to experimental results related to the study of such structure. Anomalies in water characteristics were the reason for making a suggestion that water is a mix of different types of associated molecules:
"hydrol" H_2O, "di-hydrol" $(H_2O)_2$, "tri-hydrol" $(H_2O)_3$ and so forth. Sometimes, these hypothetical associates are called polymers. The question about the maximum size for a polymer can be solved based on comparison of experimental data and theoretical results.

First of all, it is evident, that since liquid water is a fluid, capable of easy penetration of pores with diameter over 3 A, it is fundamentally different from ice and cannot have rigid crystalline structure, which is characterized by the presence of a "distant order". By the term "distant order" we mean such a way of distribution of an ordered system in space, when features of periodicity and parameters of translation symmetry don't depend on absolute shift, if such shift is a multiple of one of the spatial periods. Meanwhile, many facts suggest that water is actually characterized by a specific spatial order. Based on this, we deem it evident, that liquid water is most ordered near each of the molecules in a small volume of space. It corresponds to the fact that there must be "close order" in water, while a definite "distant order" is missing. Because of that, it would be useful to consider the function of radial distribution of the main structural elements of water (for example, nuclei of oxygen or water molecules, taken as a whole).

The simplest variant of such analysis is related to determination of the function

$$g(r) = <n_W>^{-1} dN_W(r)/dV \equiv n_W(r)/<n_W> \qquad (1.11)$$

of radial distribution of molecules of water in relation to any chosen molecule or, on the other hand, relation of local density of molecules $n_W(r)$ of water to average density $<n_W>$.

This function also determines the probability of finding a concrete quantity of molecules $N_W(r)$ in a single volume located at a distance r from a chosen molecule. Considering, that this function does, in fact, determines the correlation (mutual relation) between positions of centers of different molecules, it (in accordance with the terminology of the theory of random events) has another name – paired correlation function.

Proceeding from such definition, it is easy to find an expression for total quantity of water molecules in a spherical layer with the radius r, volume dV and width dr

$$dN_W = g(r)<n_W> dV = g(r)<n_W> 4\pi r^2 dr \qquad (1.12)$$

We can easily make some modeling estimates determining the behavior of the variable $g(r)$ at small values of r. Let's assume, that the variable r characterizes the current position of the center of a molecule. If we suppose that all water molecules are balls with diameter R, it follows directly from (1.11) that at $r < 2R$ we have $g(r) = 0$ (meaning that one molecule cannot penetrate the volume of another).

Correspondingly, with increasing r the value of $g(r)$ increases and with $r = 2R$ we have $g(r) > 1$. This result becomes immediately clear, if we use a simplified version of the model of water, which states that water molecules are in tight contact with each other. Based on the same model, with further increasing of r the value of $g(r)$ decreases and at $r = 3R$ we have $g(r) < 1$ (in an ideally periodical system $g(r) = 0$ in this case). With further increase of the distance $g(r)$ would also increase and so forth.

The quantity of water molecules corresponding to the first maximum of the probability density $g(r)$ is usually called the coordination number (and hypothetical position of these molecules is called the first coordination sphere). The same terminology is used for describing the number of molecules, whose position in space corresponds to the second, third and subsequent maximums of the function $g(r)$, which corresponds to the second, third and subsequent coordination spheres (*Fig. 1.3*).

If water structure was characterized by the distant order, the function of radial distribution $g(r)$ would be oscillating with a constant period and a constant amplitude of oscillations. However, because of unavoidable violations of periodicity with increasing distance, there is a rather fast decrease of the amplitude of oscillations $g(r)$ and with $r \gg R$ we have $g(r) = 1$. This result follows directly from *Fig. 1.3* (with a higher coordination sphere number there is a higher uncertainty about the size of its radius).

Fig. 1.3 Spatial distribution of the first three coordination spheres with maximally dense water molecules

The concept of coordination number can, strictly speaking, be applied only to crystals. Coordination numbers are just as useful for describing structures of liquids, but they have mainly statistical meaning in that case. In the case of liquids, they actually determine how close crystalline structure is to liquid structure.

In practice, the function *g(r)* is determined from the analysis of diffraction of Roentgen rays and thermal neutrons. On *Fig. 1.4* there is a paired correlation function of distribution of centers of water molecules (i.e. distribution of distances between nuclei of hydrogen atoms in different molecules of water).

It follows from this picture, that in the vicinity of any molecule of water there are three coordination spheres. Position of the first maximum and its area allow us to conclude that each molecule is surrounded (on average) by less than five other molecules located at a distance, close to the length of hydrogen links. Position of the second maximum (about 4.5 *A*) matches the length of an edge of a tetrahedron hypothetically attached to a molecule of water. This serves as one of the evidence supporting the theory that water represents a three — dimensional tetrahedron lattice consisting of interlinked molecules.

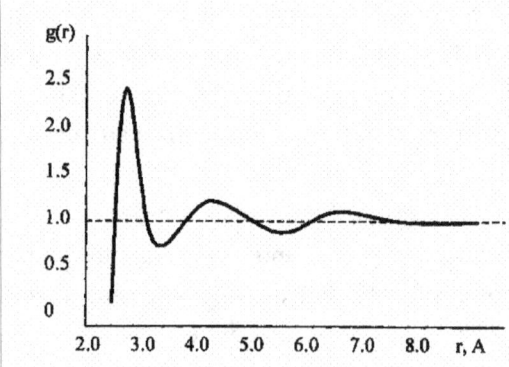

Fig. 1.4 Paired correlation function of distribution of centers of water molecules

The main question in any cohesive theory dealing with water is how well the known properties of a molecule of water (i.e. properties of its microscopic structural element) can explain all the numerous effects, which characterize the macroscopic volume of water. Due to importance of this problem it is beneficial to investigate it based on historical approach.

Let's review the evolution of understanding of the structure of water.

One of the first cohesive theories, which explained the structure of water on the basis of mathematical calculations, presumed that there is a set of similarities between properties of solid objects and liquids (quasi-crystalline theory of the liquid state). There were good reasons for such an analogy. Indeed, the density of water and the concentration of molecules in it is quite comparable with similar parameters of solid objects (including in crystals). Many liquids have the ability to preserve their form (for example, glass and sap). Thermodynamic, acoustical and electrophysical properties of many liquids are little different from the corresponding properties in solid objects. Of course, there are important differences, too. For example, ionic conductivity of electric current, non-existent in crystals, plays an important role in liquids. The diffusion coefficient in liquids is high, although, in all fairness, we should note, that coefficient of diffusion of atomic hydrogen in crystalloid palladium, for example, is so high, that it can well be compared with diffusion in some liquids.

The quasi-crystalloid theory of the liquid state was created in 1939–1954 (Zatsepina, 1998). According to that theory, the solid and the gaseous states of the matter were treated as two extreme cases of the liquid state. Within the framework of that theory, a peculiar symbiosis of two different processes – constant movement of atoms or molecules, as well as the presence of an oscillating movement around the equilibrium position – each of which characteristic either for gas or for a solid object.

Qualitatively, this model of liquid could be described the following way. Potential relief (spatial structure of potential energy for each of the molecules) of a quasi-crystalline liquid is an almost periodic three-dimensional system of potential pits and barriers. These pits and barriers are the result of self-regulated movement of all atoms and molecules. By the way, this is the approach implemented in one of the principal variants of the theory of nuclear matter (the "drop" or "hydrodynamic" model of a nucleus), where a potential barrier is created by coordinated movement of all protons and neutrons in the volume of a nucleus.

Movement of molecules, according to this model, is a combination of two independent movements – the oscillating movement in the volume of each of the potential pits and, then, a random (fluctuational) jump into a neighboring pit. During this process, the average frequency of oscillations in a potential pit is approximately the same as regular Debay frequencies in a solid object (i.e. about $\omega_D \approx 10^{13} \ s^{-1}$). Accordingly, the average duration of a jump into a neighboring pit is equal to $\tau_0 \approx 10^{-13} \ s$.

Average time of staying in the same pit $<\tau>$ is much longer than τ_0. It can be easily determined with the help of simple thermodynamic estimates, provided that in order for a molecule to leave a potential pit it must receive additional kinetic energy, exceeding pit's depth ΔW. A molecule can receive that energy only from the whole system's overall thermal energy. From the

Bolzman distribution follows, that the probability of fluctuational concentration of energy on one particle at a temperature T equals to $exp(-\Delta W/kT)$, and the probability of a molecule leaving a pit within a unit of time equals $exp(-\Delta W/kT)/\tau_0$. Hence, we find that the average time of a molecule's stay in one potential pit equals

$$<\tau> = \tau_0 \, exp(\Delta W/kT)$$

Since $\Delta W >> kT$, $<\tau>$ turns out to be much larger then τ_0.

By the same logic, we can find an approximate expression for the average speed of movement of a molecule in a liquid. Assuming that the average distance between centers of neighboring potential pits equals $2R$, average speed is

$$<v> \approx 2R/<\tau> = (2R/\tau_0) \, exp(-\Delta W/kT)$$

This expression agrees very well with the experimental data obtained from studying diffusion and ionic electric conductivity in water. In addition, it's obvious that such a simplified model of liquid is based on very rough assumptions and cannot explain many facts observed in experiments (in particular, its irregular compressibility and "water's memory").

Another approach to a model of liquid as an analogue of solid objects was made in the works of Bernal (1959, 1960). Based on specific rules of symmetry, he came to the conclusion that correlation features in distribution of water molecules neatly match the experimental data provided that spatial distribution of molecules of water is characterized by pentagonal symmetry. This model had allowed Bernal to explain accurately such anomalies in simple liquids as their capacity for excess chilling without a transition to the solid state, high entropy of water, its liquidity and several other effects.

The "dodecahedron model" of L.Pauling (1959) holds a special place among numerous models of water structure. In the basis of this model there is the idea that hydrates of gases can create spatial tetrahedral frames with vary large (in the scale of a water molecule) internal micro cavities with rigid atomic walls. Such systems have been dubbed clathratic hydrates. The main elements of this structure are right polyhedrons (dodecahedrons), each one of which is interlinked by hydrogen links. Each one of them has 12 pentagonal sides, 30 edges connecting those sides and 20 ends with 3 edges converging in each one of them. Water molecules are located in the joints of this spatial frame. If there are alien dissolved gases in water, they are located in internal micro cavities. In the Pauling model, frame molecules of water form the walls of internal micro cavities, each one of which can be characterized by an inscribed sphere with diameter 5.2 A. Molecules of methane CH_4, oxygen O_2, nitrogen N_2, chlorine Cl_2 may be located in the volume of micro cavities. Micro cavities are connected with external environment by windows with diameter about 2.5 A, which is little less than diameter of a water molecule (≈ 2.76 A). These micro cavities are separated from the "outside world", i.e. from the main volume of water, by a certain potential barrier.

We can easily make sure, that the internal size of these micro cavities is much larger than the diameter of a water molecule. Therefore, molecules of water may be located in those cavities without forming any hydrogen links! Moreover, due to high symmetry of the electrostatic field inside micro cavities there is a certain limitation on formation of hydrogen links for such molecules of water, which makes them hydrophobic. Calculations have shown, that the density of such structure (without filling it with water molecules) equals 0.80 g/sm^3. If micro cavities are filled with water molecules, density is close to normal water density 1 g/sm^3.

Due to the named circumstances, existence of micro cavities not filled with water is possible, along with a stable and isolated from the outside environment condition of water in those micro cavities.

Minute calculations show, that within the temperature range from 0° to 30°C the Pauling model can sufficiently explain all features of water (including its unusual compressibility with a minimum at the temperature of 4°C).

It will be shown in chapter 2 that many anomalous features of water (even the most important one of them – the –"water memory") can be successfully explained on the basis of the Pauling model in its comprehensive application.

In his works, Frank studied the possibility of uniting the clathrate Pauling model with the cluster model. In that case, the dodecahedron frames can sometimes connect with each other by hydrogen links and, thus, form groups with ordered structure (i.e. clusters). Since there is a very strong correlation between close hydrogen links, appearance and removal of hydrogen links takes place in a coordinated fashion, being synchronized in time. Such nature of the link allows us to suggest that "flickering clusters" appear and disappear in water. The mechanism of forming such clusters is the following.

At the room temperature more than half of hydrogen links in water is severed. Nevertheless, an even distribution of severed links over the molecules is not advantageous due to collective (cooperative) character of their generation. Remaining links are redistributed in a way that allows achieving their maximum concentration forming associates with the maximum number of links per molecule. It corresponds to formation of a structure, close to the structure of ice, providing the maximum number of links. These associates are rather unstable. Local fluctuations of energy lead to their disintegration and appearance of new ordered associates – clusters. The life time of clusters – about 10^{-10} s, or about 1000 molecular oscillations.

Later, Frank and Quist (1961) as well as Frank and Wen (1957) both considered another variant of a double structured cluster model of water. They studied water as a balanced mix of monomeric molecules and molecules, belonging to ice remains. In this model, great significance is given to a collective mechanism of appearance and disappearance of hydrogen links. In liquid water the "flickering clusters" with ordered internal hydrogen connection (there are ice-like groups of water molecules in those clusters) alternate with vast areas, where hydrogen links don't exist.

Various aspects of this problem are actively discussed at the present time. In one of the most comprehensive reviews of late (Buljonkov, 1991) verified data related to the tetrahedron location of hydrogen links is used to create a universal clathrate of water consisting of three hexacycles. The work also studies correlations between the spatial structure of the frame formed by molecules of water and the structure of proteins and nucleotides of DNA. According to the author's calculations, each pair of nucleotides in DNA has no less than 16-28 water molecules linked with it. It is also shown, that spirals of water triplets may generate in ionic channels of biological membranes. Each of these triplets contains 20 molecules of H_2O. Ions of potassium, sodium and magnesium can pass inside these channels. Features of interaction of water with surfaces of biological macromolecules will be discussed in chapter 2.

Another work (Davenas, 1988) discovered the presence of certain features suggesting activity of some biologically active substances at an extremely high degree of their dissolution (in homoeopathic doses). It turned out, that such solutions have a real effect. For example, they cause degranulation of basophiles, which could be explained only by existence of quasi stable forms of structured water. Similar results were presented in 1994 by Polak. He also showed that in electrochemically-activated water there are three structurally separated meta stable forms with different times of relaxation.

In recent years, with progress in computer technology, purely quantitative methods of water structure analysis are being developed.

The conceptual part of such studies is based on the continual model of water, which postulates the presence of an uninterrupted net of hydrogen links. Parameters of the links are calculated using mathematical and statistical methods (for example, the Monte-Carlo method). These methods, when used correctly, make it possible to determine the spatial location of hydrogen links.

The most conventional application of such studies includes the study of behavior of a relatively big system of molecules, each one of which nay be presented in the form of a certain symmetrical object, like sphere, which are connected by four hydrogen links. Since modern computers allow to make calculations for several hundreds of interacting molecules, such a system has, largely, the features of a real macrosystem of water. In particular, it was established by calculations that the quantity of molecules H_2O not participating in formation of any hydrogen links does not exceed 10%. Such model based on quantitative modeling may be used for the study of thermodynamic, acoustical, optical and other properties of water. Among the quantitative methods, the Monte-Carlo method is used most often. It can be simply explained by the following.

First, the initial configuration of molecules is investigated. Normally, it corresponds to homogeneous (by volume) distribution. For this system, the initial value of the system's energy is calculated. Then, there is a trial shift and turning of one of the molecules. A new value for energy is calculated for this new deformed configuration. Different shifts and turns of the same molecule provide different values for total energy. When a configuration with minimum energy is found, it is taken for new initial condition. Then, the same procedure is repeated for the next molecule. The order of disturbing molecules is obtained at random, which is the reason for calling this method the Monte-Carlo experiment. By repeating the experiment for different variants one can find such a configuration, which determines the absolute minimum for energy of the whole system (when any shift or turn of any molecule leads to increasing energy). Such optimized configuration reflects the state of thermodynamic balance of the entire volume of water.

Based on the analysis of calculations conducted earlier, we can note, that the results of numeric modeling obtained on such a model with three-dimensional net of hydrogen links match ordinary water with permissible accuracy.

Another aspect of structural modeling of water is related to the study of its behavior in limited and small volumes. This task is especially important in its implementation for biological systems, each one of which is limited by definition. The presence of borders causes principally unavoidable heterogeneity and flawlessness of a system, since a border is always the main defect. In biological macromolecules and their assemblies there is often the effect of reduction of dimension of the water structure – instead of the three-dimensional system in an unlimited volume of water we have to consider two-dimensional water (for example, on the inner surface of ionic membranes) or even one-dimensional (in very fine channels). Numeric methods are used very successfully in these models.

The question of which of the models is most probable still awaits for its final answer. However, we can note with confidence that the "true" model will be, most likely, a combination of the coherent and clathrate models. Areas with ordered structure are, in fact, often found in water and their share in the volume of water reaches 5–10%. Each one of such areas contain several tens of molecules H_2O, while their size equals 20 A. Using the terminology of the clathrate model, discussed above, these areas are analogues of an ice-like micro phase. The average distance between such areas is equal, approximately, 50 A (Sokolskii, 1990).

1.5   INFLUENCE OF IMPURITIES ON THE STRUCTURE OF WATER

The futures discussed above relate to absolutely pure water, without any impurities. However, such water doesn't exist!

First of all, water, having a very large static dielectric permittivity, can weaken interaction between charges in any objects found in its volume. This circumstance immediately leads to a possibility of separation of charges with their placement in volume of water. Water is the most universal solvent. It is capable of absorbing materials, from which the walls of a vessel containing it are made. It can extract gases from air and absorb them. Moreover, even absolutely clean water, by participating in the process of radiolysis under the impact of natural radioactive background, can enrich its volume with numerous free radicals. The pattern of water radiolysis has been discussed above.

Impurities strongly affect the structure and characteristics of water.

Samoilov (1957) has demonstrated that the structure of water depends heavily on interaction of its molecules with ions. Clearly, the character of such interaction depends on many factors.

Geometric compatibility of the size of ions in lattice of ice has big significance affecting the structure of water. For example, an effect of a small ion, inserted into the spatial structure of a water molecule would be different from an effect produced by a large ion.

Another important factor is the sign of a charge of an ion. The most prevalent anions (negatively charged ions) normally have large size and do not have a significant hydrate shell (ions OH^- are the exception from that rule). Because of that, interaction of anions with water is reminiscent of interaction of molecules of water between each other.

The energy of interaction of cations with molecules of water by far exceeds the energy of interaction of molecules between each other. Situation, when a hydrate shell consisting of a large quantity of oriented water molecules is quickly formed near each ion is typical. Meanwhile, ions can form small compounds. In the first approximation, the hydrate shell is characterized by a position of water molecules in the first hydrate layer. Due to reorientation of molecules, there emerges a new structure of water in the immediate proximity to an ion (*Fig. 1.5*).

Another important question is related to the lifetime of a hydrate system. Position of water molecules in it is described by how much does their orientation contributes to formation of hydrogen links. A molecule of water in the first hydrate layer stays in stable condition for much longer than a molecule in the volume of homogeneous water. For example, a molecule H_2O, linked with an ion of cobalt Co^{2+}, is in direct contact with this ion for about 10^{-7} s, which is 10000 times longer than typical time of connection of water molecules between each other (Antonchenko, 1989).

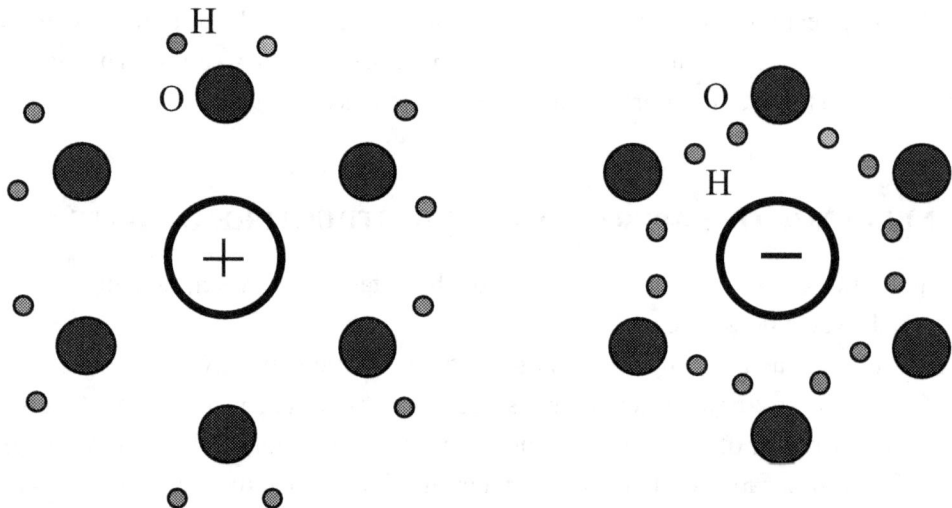

Fig 1.5. Structure of the first hydrate layer in water for cations and anions

Next to the hydrated water layer, lies a layer of water, which structure is damaged due to mutual influence of external (distanced from an ion) molecules of water and orientating influence of an ion. Because of a large radius of this shell, its volume exceeds the volume of the hydrated water layer. Further away is water with an undamaged structure.

Experiments involving the use of numeric modeling based on the Monte-Carlo method have shown that appearance of ions in water affects the dynamics of movement of water molecules and, therefore, the overall structure of water.

There are two opposing tendencies, which define this process.

On the one hand, since the process of hydration causes distortions of homogeneity of water and its falling out of order, it means an increase in entropy. This tendency is very sensitive to water temperature and its manifestation constitutes the main contribution to changing of the balanced state of water at a low temperature. With a higher temperature, the magnitude of this tendency decreases, because thermal movement of molecules becomes more significant. On the other hand, reflected electrostatic field of an ion makes an orienting influence on molecules of water, since it polarizes them. Such orientation corresponds to lower entropy.

In the end, with higher temperatures, micro heterogeneities in water are destroyed while transferring movement is increased.

Sometimes, the hydrate shell of an ion can assume the form of simple molecules.

For example, proton H^+ formed as a result of radiolysis can unite in water with a molecule of water, forming a complex H_3O^+. Ions H_3O^+ and O^- unite three molecules of water each in their proximity, creating a structure $H_3O^+(H_2O)_3$ and $OH^-(H_2O)_3$. Existence of such ordered systems is confirmed by various methods (Brady, 1957, 1960).

It is also worth noting another condition: as was said before, ions H^+ and OH^- are not localized in space, but are moved by a migration mechanism at a high speed. The speed of their movement is so great that, actually, an ordered effect of ions H^+ and OH^- is brought to average by all molecules of water, thus producing the effect of remote action in bringing a structure to order.

Based on the analysis of experiments (Samoilov, 1957) it was established that the phenomenon of hydration can lead to less mobility of molecules of water near each ion than in clear water, although they can also be more mobile. The former effect was called positive hydration, while the latter – negative hydration. The type of hydration depends on the nature of ions. In particular, cations of alkaline and alkaline-earth ions, as well as anions of chlorine, bromine and iodine have poor connection with molecules of water. This leads to negative hydration (Sokolskii, 1990).

The process of "distant" interaction of ions, which occurs during the process of polarization of intermediate water molecules, is very important. This process has a simple explanation. If a cation polarizes an adjacent water molecule in the course of such interaction, nuclei of hydrogen contained in it become more positive and acquire acidic properties. Such atoms, when compared to atoms of "regular" hydrogen in non-polarized molecules H_2O, create hydrogen links with anions much easier. Bigger radius of operation of an electrostatic field of an ion's charge compared with a field of an induced dipole also supports wide radius of interaction. All these factors produce the effect of dynamic orientation of water molecules in moderately concentrated solutions. In the process of such dynamic orientation each ion becomes surrounded by a more "ordered" water in a direction which connects it with an oppositely charged ion. And vice versa, the hydrated shell becomes "disordered" in a direction of an ion of the same charge sign (Antonchenko, 1989).

Another important effect is related to the situation, when under the influence of ions dissolved in water concentration of hydrogen ions changes. This process is called hydrolysis and it is similar – in its resulting effect – to the effect of radiolysis considered above. The main mechanism of hydrolysis for cations is based on the fact that strengthening of a covalent link between a positive ion and a molecule of water causes deformation of the latter and, ultimately, a possibility of generating an ion of hydroxonium H_3O^+. It results in lowering of pH of a solution, which assumes acquires acidic properties.

For anions, the mechanism of hydrolysis is different, but the result is the same. The closer the affinity of an anion to hydrogen, the stronger the hydrogen link is. Meanwhile, a proton of a water molecule shifts towards an affecting ion, which immediately causes weakening of the main (for a water molecule) link along the O-H line. Reduction of this link leads to a possibility of tearing off an ion of hydroxyl OH^- from the water molecule H_2O. In order to do that, an energy ΔW_{OH-} of generating an ion OH^-, which becomes less with deformation of a molecule, discussed earlier, needs to be applied. Probability of such a process is proportional to the value $exp(-\Delta W_{OH-} /k_B T)$ and it sharply increases with higher temperature. Generation of additional ions OH^- reduces the pH value for a solution, while it acquires, as in the case with cations, acidic properties (Antonchenko, 1989).

Dissolved gases exert big influence on the structure and properties of water. It is conditional on the presence of micro cavities in the structure of frame molecules in the Pauling model. Depending on the type of molecules of gas, their inculcation into the inner frame micro cavities can either stabilize that frame or make it less stable. It has been estimated, for example, that when atoms of an inert gas are dissolved in water, microscopic objects like ice as well as other structures with bent hydrogen links may form around them.

1.6   INVESTIGATING BIOPHYSICAL FEATURES OF WATER AND WATER MEMORY

We can make now make some preliminary conclusions about our analysis of the features of water and its structural models. Considering the role of water in all vital processes, from geopolitical perspective, possessing water sources and technologies for its treatment with the purpose of achieving optimal biological condition should be perceived in the same context with competition for energy and food resources.

Studying physical features of water, its structure, changing of its characteristics under the influence of various physical factors and its own influence on conditions of biological objects is one of the most complicated and complex problems of modern natural science. In physical and quantum chemistry water is regarded as an extremely complex substance. The difficulty of studying water is due to the fact that, being a mixture of various dynamically changing structures (clusters, associates) water constantly changes its characteristics. It possesses a number of paradoxical physical qualities in comparison with other substances. It is extremely sensitive to the impact of physical factors – temperature, magnetic and electric field, mechanical effects. Moreover, it has "memory" for past physical effects and conditions lasting for relatively long periods of time (in the order of days) when it preserves altered physical and chemical qualities.

The role of water and a very significant alteration of its structures and physical characteristics in biological systems is the subject of intense biophysical research.

Currently, there are different theories and models explaining a number of features and structures of water and its physical qualities. Problems related to very weak energy and information influences on water and biological objects are still unsolved, since water is the main intermediary in changing qualities of biological systems. Similarly unanswered are questions pertaining to effects of homoeopathic drugs using practically clean water for achieving directed effect on an organism when water preserves informational "traces" of substances contained in it before numerous dilutions.

Some of these problems will be considered in chapter 2.

One of the main peculiar features of water is its ability to "memorize" structural changes and preserve that information for a long time.

We can offer many examples of such memory.

In 1933, Bernal and Fowler reported sudden changes of viscosity and surface tension of water accompanying increasing temperatures.

In his work, Sikorskii (1959) reported that after melting twice-distilled water (bi-distillate) and its subsequent holding at a constant low temperature, its dielectric permittivity slowly increases reaching its normal value of $\varepsilon \approx 90$ only after 800 seconds. In order to exclude the influence of tiny pieces of ice in this process, which could have remained in water after melting, an alternative experiment was conducted, when water with the same temperature was obtained not by melting ice, but by condensing steam. It turned out, that density of such water immediately after condensation and cooling was higher than the equilibrium value at that temperature. Only after 1800 seconds its density had reached 1 g/sm^3 (Klassen, 1973). Along with that, relaxation (self-induced change in time with an eventual establishment of a stationary condition) of water absorption in the infrared part of the specter was also noticed. Since infrared specter reflects profound alterations in an object, the latter circumstance give reasons to believe that such relaxation had reflected the process of creation of a spatial global structure of water. Considering many similar unusual results, we can assume that water possesses a certain "structural memory". That memory may be unrelated to the fact that water actually changes its volume structure in a large spatial interval. That could also be local structural changes or it could be changes caused by redistribution of ions and molecules. The problem of fixing the structural condition of water stems from the fact that the lifetime of atomic-molecular complexes with hydrogen links known from quantum chemistry is very short. Obviously, these are not distributions caused by "normal" changes of conditions of molecules in a volume of water, which preserve their stability for not more than 10^{-10} s (in a homogenous volume of water) or 10^{-8} s for the case of hydrated water. With such standard treatment, it could be expected that all types of associates must rapidly exchange water molecules or protons, which wouldn't allow us to speak of a fixed quasi-stationary structure of water, let alone its memory. But the experiments tell the opposite – long-term structural configurations in water really exist. Actual memory of water is preserved for the duration of several hours or even days! This is the paradox, which has to be resolved. One of the potential mechanisms of long-term memory of water will be discussed in chapter 2.

In summary, we can confidently say that water structure determines, apparently, most of its unusual properties.

Conclusions and future goals:

1. Despite a huge number of conducted experiments, water and its solutions remain, in many ways, inadequately researched systems, particularly in the area of directed effect on condition of biological objects.

2. There have been certain achievements in the area of creating theoretical understanding of the structure of water including as part of live objects. To a lesser degree, it concerns determination of mechanisms of influence of magnetic and electrostatic fields on water.

3. Existing model-based concepts of the structure of water and its solutions are quite complicated. So far, there have been no model suitable for description and forecasting of processes occurring in water under the influenced of a wide range of real physical factors. The description of physical and chemical effects on water in many models is mainly qualitative.

4. Reliable mathematical modeling of processes in water environments is hindered due to extremely big sizes of models used for imitation of water structures and dynamic processes taking place there.

5. Precise identification of mechanisms and modeling of cooperative, structuring Van Der Waals forces affecting macro molecules of biosystems in water environments presents a separate unsolved problem. These forces are responsible for cellular interactions, creation and functioning of cellular membranes, tissues and organs.

6. Considering vagueness of many mechanisms of influence of the structure and physical characteristics of water on condition of biological systems, it would be helpful, at the present time, to emphasize experimental research with the purpose of determining the patterns of influence of various physical factors on water and, indirectly, on biological systems, making attempts to summarize and organize the results on the basis of the existing qualitative models.

7. The role of water in biological systems subject to external influence (in particular, effects of ionizing and non-ionizing fields) is not adequately researched. Using the traditional "target" method immediately leads to a barrierless concept of radiobiology, which doesn't correspond to many experimental facts. This concept conflicts with the phenomenon of a beneficial influence of small doses of ionizing radiation – "hormesis".

8. Energy-informational effects on water accompanied by changes in physical and biological properties of water, which was in contact with specially treated water are of special interest. Also of special interest is research of possibilities and consequences of negative effects on water put in contact with certain devices, as well as devices, which remove such possible negative effects.

9. The study of minute energy-informational effects on water is quite science-intensive, complicated and require a high degree of professionalism along with special laboratory and and equipment conditions for conducting physical and biological experiments on water and biological objects.

Literature to INTRODUCTION

Amiragova M.I., Duzhenkova M.A., Savich A.V., Shalnov M.I. "Primary Radiobiological Processes." Atomizdat-Publishers, Moscow, 1964

Antonchenko V.I. Physics of water, Kiev, 1986 (in Russian)

Bell S. // Jour. Mol. Spectrosc. v. 16 (1965) *p.205*

Bernal J.D. // Nature, v. 183 (1959) *p. 141*

Bernal J.D. // Nature, v. 185 (1960) *p. 68*

Bernal J.D., Fowler R.H. // J.Chem. Phys., v. 1 (1933) *p. 515*

Brady G.W. // J. Chem. Phys., v. 33 (1960) *p. 1079.*

Brady G.W., Krause J.T. // J. Chem. Phys., v. 27 (1957) *p. 304*

Buljonkov K.A. // Biophysics, v. 36 (1991) *p. 181* (in Russian)

Coulson C.A., Glaeser R.M. // rans. Faraday Soc., v. 61 (1965) *p. 380*

Dahl L.W., Anderson H.C. // J. Chem. Phys., v. 78 (1983) *p. 1980*

Davenas E., Beauvais F., Amara J. et. al.//Nature, v. 333 (1988) *p.816*

Davis C.M, Litovitz T.A. // J. Chem. Phys., v. 42 (1966) *p. 2563*

Frank H.S., Quist A.S., J. Chem. Phys., v. 34 (1961) *p. 604*

Frank H.S., Wen W.Y. // Discuss. Faraday Soc., v. 24 (1957) *p.133*

Frank H.S., Wen W.Y. // Discuss. Faraday Soc. v. 24 (1957) *p. 133*

Klassen V.I. Water and magnet, Nauka, M., 1973 (in Russian)

Koulson Ch. Valence, Moscow, 1965 (in Russian)

Krasnov et. al. Stable molecular non-organic compounds, Leningrad, 1968 (in Russian)

Lennarri-Jones J., Pople J. A. Proc. Roy. Soc. // A202, 166, 1950.

Nemetry G., Sheraga H.A. // J. Chem. Phys., v. 36 (1962) *p. 3382*

Pauling E. Hydrogen bonding // Ed. D.Hadzi, London, Pergamon Press, 1959

Physical Constants. Ed. Grigoreva I.S., Meilikhova E.Z., Moscow, 1991.

Polak E.A. // Biophysics, v. 36 (1994) *p.565* (in Russian)

Pople J. A. Proc. Roy. Soc. // A202, 323, 1950.

Pople J.A. // Proc. Roy. Soc. London, v. A205 (1951) *p. 163*

Samoilov O.I. Structure of water solutions and hydration of ions, Moscow, 1957 (in Russian)

Sceats M.G., Rice S.A. // J. Chem. Phys., v. 72 (1980) *p. 3226*

Sikorsky Yu.A., Vertepnaya G.I., Krasilnik M.G. // Izvestiya vuzov. Phizika, 1959, No 3, C. 76 (in Russian)

Sokolskii Yu.M., Magnetized water, Moscow, Chemical Publ. House, 1990 (in Russian)

Vysotskii V.I., Pinchuk A.A., Kornilova A.A., Samoylenko I.I. // Radiation Biology. Radio-ecology, v. 37 (1997) P. 494-507.

Yukhnevich G.V. Infra-Red Sprectroscopy of Water, Nauka - Publishers, Moscow, 1973 (in Russian)

Zatsepina G.N. Physical properties and the structure of water, Moscow State Univ. Publ. House, 1998 (in Russian)

2. WATER AND ELECTROMAGNETIC FIELD AS UNIFIED SPACE IN TIME. WATER MEMORY AND FEATURES OF WATER INTERACTION AND ELECTROMAGNETIC FIELD

2.1 SPATIAL STRUCTURES AND THE PROBLEM OF LONG-TERM WATER MEMORY

Water has a number of unique features, among which there is its long-term "memory." Numerous experiments confirm the existence of water memory, which is activated under the influence of some physical fields (for example, magnetic field, mechanical impact, abrupt temperature or pressure change) and may store information about such influence for many hours and days.

For instance, according to the data (Klassen, 1973) in water subjected to a constant magnetic field changes of all the main characteristics are observed. In particular, during a change of the field strength from $H = 1900$ Oersted to $H = 5700$ Oersted, the value of pH of a chemically pure water (redistilled) changes by 5.1–9.1%, while its surface tension – by 2.2–7.3%. Such magnetic treatment alters the specter of infrared absorption of water. In water put through an area of a strong enough magnetic field, the efficiency of dehydration processes of dissolved diamagnetic ions is reduced and, accordingly, the efficiency of dehydration of paramagnetic ions is increased.

In the same water the surface wettability also considerably changes. The effect turns out to be quite ambiguous – if a surface contains silicon, the wettability improves, while without silicon it usually decreases.

Magnetic treatment of water leads to a very significant change in the rate of dissolution of many salts. For example, under the impact of a strong magnetic field on water conducted in the mode of a sharp change of direction of that field the rate of dissolution of sulfuric magnesium increases by 120 times.

Results of some works suggest that under the influence of a weak microwave irradiation there is a significant change of index of refraction of water solutions of proteins contained in blood plasma. Further experiments have shown that a change of index of refraction of water itself contributes the most to that outcome.

The presented anomalous features of water subjected to the effect of a magnetic field remain for many hours and days.

The experiments show that such activated water has altered physical and chemical features and in some cases may have a special effect on biological objects (including medicating effects during treatment of some illnesses).

Another aspect related to the problem of water memory is due to the possibility of preserving information in water about dissolved chemical compounds at a very high degree of dissolution. In fact, such problem corresponds to the field of classical homoeopathy.

Among the works causing big resonance we would point at one published in the reputable magazine *Nature* (Davenas, 1988). In that work there was identified preservation of information in water about trace quantities of some biologically active substances (i.e. its factual chemical activation). That information was preserved even after an ultimately strong dissolution, when molecules of a dissolved substance in water were completely absent. This work describes investigations in classical immunology conducted by the scientific group of a French biochemist J. Benveniste. He studied the effect on blood cells, called basofilles, protein molecules specifically affecting these cells and causing their specific response reaction, which is called degranulation. Conventional biochemistry is based upon the belief that the higher the concentration of such proteins the higher the rate of such reactions. Accordingly, with lower concentration the rate should also be lower. In contrast to that, the experiments have revealed that at even the strongest dissolution of protein molecules (antipolyglobulines) when their relative concentration was lower than 10^{-30} (which is equivalent to only one molecule per 70 litres of water!) a clearly pronounced effect of degranulation of basofilles was observed. Since the volume of a pan where the experiments were carried out was naturally much lower than 70 liters, it means that there was not a single protein molecule in the volume of water after repeated dissolutions.

The experiments, conducted during that project have shown that this information may be preserved in undisturbed water at an adequately low temperature but it may also be effectively "erased" during such classical types of influences as the impact of ultrasound, strong heating or phase transfer during freezing and subsequent melting of ice.

Note, that the medical aspect of influence of activated water is studied very poorly, although its influence has been proved in many experiments.

As an example of the influence of activated water on simplest microorganisms we can give the results of studies performed at the department of Biology of the Moscow University (these unpublished results are provided in the dissertation work of Zenin (1999)).

The process of water activation consisted of its treatment by alternating magnetic field created by a standard magnetic mixer for 7 and 15 minutes. Such water goes through a series of physical and chemical changes (including different specific conductivity). In particular, preliminary studies have demonstrated that in the result of treatment of distilled water for 7 minutes its specific conductivity had increased by no less than 10–15 times. The law of increasing conductivity was close to linear. After ceasing of the impact of alternating magnetic field there was a slow relaxation of conductivity until its lowering to the initial value within 20–25 minutes. A completely different situation has had place after a longer treatment of water. At the initial stage (during the time of influence of alternating magnetic field) there was an analogous increase of water conductivity, while after magnetic influence was stopped, there was an effect of a spontaneous additional increase of water conductivity by 2.5–3 times within 15–20 minutes instead of reduction of conductivity. Such result is attributed to the feature that the system of water memory is characterized by a threshold time of irreversible activation within the time range from 7 to 15 minutes.

That activated water was used for the study of the effect on the *spirostomes*, one of the types of infusoriums, immediately after the activation process.

In the case of water subjected to a short-term below-threshold influence of a magnetic field, there was the effect of a short-term inhibition of the movement activity, which disappeared after 30–40 minutes.

In water subjected to a long-term (above-threshold) activation, an irreversible paralysis of the *spirostomes* with full suppression of all signs of activity was noted.

These facts constitute one visible side of the problem. That side attracts the most interest, but, at the same time, it causes most objections.

The other side of the problem is based on explanation of these effects and determination of their mechanisms. It is mostly important for scientific opponents. Let's investigate it closer.

At the first glance, it appears that water as a specific physico-molecular object cannot have any long-term memory. It follows from simple estimates.

For a long time continuous (quasi crystalline) model of water was the dominant one. Within the framework of this model the spatial structure of potential energy for each one of H_2O molecules is nearly a periodical three-dimensional system of pits and barriers. This relief is the result of a self-regulating movement of all water molecules, which represents a combination of two independent processes – vibration movement in each one of potential pits and random (fluctuation) leap into a neighboring pit. The average frequency of vibrations in potential pits is approximately the same as the Debay frequency in a solid body (about $\omega_D \approx 10^{13}\ c^{-1}$). The average duration of a leap into a neighboring potential pit is equal $\tau_0 \approx 10^{-13}\ c$. Average time of staying in one pit

$$<\tau> = \tau_0\ exp(\Delta W/kT) \approx 10^{-9} - 10^{-10}\ c \qquad (2.1)$$

is determined by the water temperature T and the energy of activation $\Delta W \approx 0.2\ eV$ of the diffusion process (the height of the barrier between neighboring pits). Staying within the framework of this model it is easy to make a conclusion that water memory must be preserved for not much longer than the value $<\tau>$, which is by many orders less than given numerous experiments.

There can only two ways out of this logical dead end – either the experiments are not reliable, or the continuous model is incomplete (or wrong).

Continuously increasing number of reliable experiments shows that the continuous model is inadequate for describing the water structure. Moreover, continuous non-structural water (according to the data by the Nobel Prize winner Juan Lee) should be not a liquid but gas because of a relative weakness of hydrogen links!

The presence of a spatial structure in a volume of water was first proved by Bernal (1933).

Calculations made on the basis of quantum chemistry have shown that water molecules participate in creation of molecular assemblies and may form different types of associated molecules: "hydrol" H_2O, "dehydrol" $(H_2O)_2$, "trihydrol" $(H_2O)_3$ and so on. Further studies have shown that even much larger associates (clusters) may form in water from water molecules, whose structure resembles small pieces of ice. As a rule, these clusters are unstable and appear and disappear spontaneously. The dynamics of such associates lies in the basis of the cluster model of water (Nemethy, 1962).

More detailed studies (for example (Samoilov, 1957)) have shown that the so-called "clathrate" model is the one closest to reality. In its final form this model was developed by Pauling (Pauling, 1959). In the basis of the Pauling model there is the concept that unification of atoms of oxygen and hydrogen can create spatial flexible tetrahedral frames.

Formation of a tetrahedral frame was due to the fact that the natural spatial angle between OH-links in a free water molecule H_2O is equal to 104.5°, which is sufficiently close to the exact value of the tetrahedral angle 108°. In order to achieve an additional angle of 3.5° in this link an insignificant energy would be required, while the existence of an additional curve would considerably increase rigidity of the crystalline frame (a similar situation occurs, for example, in such a purely construction material as preliminary stressed concrete).

In the joints of the crystalline frame there are very large (in the scale of a water molecule) micro cavities (microscopic empty spaces) with rigid atomic walls. The main elements of this structure are right polyhedrons linked to each other – dodecahedrons. Such systems are called "clathrate hydrates". The entire frame is held together by hydrogen links. They fasten together a system of pentagonal dodecahedronic polyhedrons from ions of oxygen and hydrogen, which form the walls of the micro cavities. Each one of the polyhedrons may be characterized by an inscribed sphere with radius about $R_c \approx 2.6$ A. All polyhedrons have 12 pentagonal facets, 30 edges connecting these facets and 20 vertexes with 3 edges converging in each one of them. On the vertexes of these polyhedrons there are 20 molecules of water H_2O, each one of which having three hydrogen links. According to the available data (Zenin, 1999) any 3 polyhedrons may be unified into stable associates containing 57 molecules of water. Out of these 57 molecules 17 have fully saturated hydrogen links and they form a tetrahedral hydrophobic central frame, while in 4 dodecahedrons there are 10 centers of formation of hydrogen links (O-H or O) located on the surface of each one. The space structure of the system of clathrate hydrates in water is presented in *Fig. 2.1*.

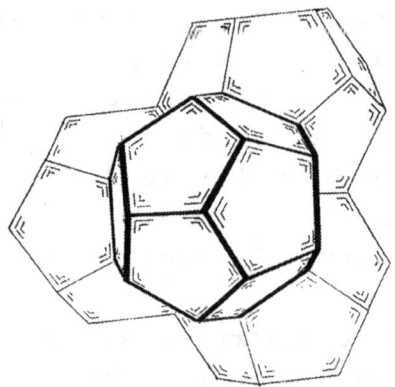

Fig.2.1 The system of clathrate hydrates in water

Beyond this frame there are quasi free molecules of "regular" isotropic water, the features and the structure of which approximately matches the continuous model. Micro cavities are linked to the outer space by windows with diameter of about 2.5 A, which is slightly less than diameter of a water molecule ($2R \approx 2.76\ A$). In the result, each of the micro cavities is separated from "external" amorphous quasi free water by a circular potential barrier with width about 0.13–0.15 A bounding each window. Relative quantity of molecules of "frame" water at room temperature is 20–30% increasing with lower temperatures. In the volume of micro cavities one molecule of H_2O, CH_4, O_2 or N_2 may be accommodated.

Due to the presence of a strong and symmetrical (relative the center of micro cavities) electrostatic field there is a certain ban on formation of hydrogen links of water molecules with the walls inside the micro cavities. In this case there is such a non-trivial phenomenon as repulsion of free water molecules from the walls of the frame also consisting of water molecules (in other words, water molecules in the volume of water become hydrophobic)! The average density of a clathrate frame (without filling it up with water molecules) equals 0.80 g/sm^3, i.e. the micro cavities occupy 20% of the full volume of a structured water frame. If the micro cavities are saturated with molecules of water the density of water would be close to 1 g/sm^3.

The results of direct measurements (Zenin, 1999) have shown that optical properties of structured and amorphous water at the same temperature differ by a very large margin. Specifically, the difference in characteristics for refraction of the clathrate frame and amorphous water reaches in some cases 4–5%, which testifies of spatial ordering of the clathrate frame.

The Pauling's clathrate model explains very well all features of water (including its anomalous condensability). The DNA structure ideally matches the spatial structure of such framed water, given that each macromolecule of DNA regulates water at the distance of up to 300–500 A away from its surface. Many works address the possibility of unification of the Pauling's model with the cluster model. In that case, separate elements of clathrate frames may, from time to time connect with each other by hydrogen links and form groups with ordered structure (or clusters). Since there is a strong interdependence between adjacent hydrogen links, creation and elimination of hydrogen links happens in a correlated manner synchronized in time. Such character of a link allows us to say that there exist appearing and disappearing "flickering clusters" in water. The lifetime of such clusters is about 10^{-10} s or about 1000 molecular vibrations.

2.2 HIERARCHY OF THE ELEMENTS OF WATER STRUCTURE AND MOLECULAR SYSTEM OF LONG-TERM WATER MEMORY

The examined features of spatial water structure show that water molecules are always distributed between two loosely connected systems: the quasi amorphous non-structured water and the quasi crystalline structured system of clathrate hydrates. During the process of external influence on water (i.e. water activation) there is a significant change of its structure and parameters.

Proceeding from the scale and mechanism of activation it is possible to single out two different hierarchical levels of organization for water structure (the macro level and the micro level).

The first hierarchical level (macro level) of water structure relates to the global spatial structure of water and determines the shape and location of its spatial frame. This level is characterized by the presence of a system of clathrate hydrates, which form stable dodecahedronic polyhedrons from ions of oxygen and hydrogen. Inside the volume of each of these polyhedrons there are void micro cavities with solid walls. With the help of stable hydrogen links dodecahedronic polyhedrons are connected into associates, which may be united into large associates (macro clusters). In the space between the macro clusters there is quasi-amorphous water. Therefore, the macro level of structural organization of water corresponds to a balanced distribution between the amorphous water phase and another phase of water, represented as a system of macro clusters. With an impact of external parameters this distribution may change. For example, with lower temperatures the volume of macro clusters increases while the volume of quasi-amorphous water decreases. With growing temperatures the volume of macro clusters is reduced and, above that, each one of them may be divided into several smaller parts. Meanwhile, the volume of quasi-amorphous water, naturally, increases. The same changes may take place during other types of influences (for example, under the effect of ultrasound on water environment). Due to a strong dependency from external influences the macro level of water structure is not efficient enough for sustaining a system of water memory, resistant to external destructive influences. Meanwhile, it is obvious that in the absence of very strong destructive influences the process of recording of information in the form of a system of specifically arranged clathrate cells is quite possible. Simply stating, separate polyhedrons of the clathrate frame may be connected with each other by several alternative ways. One important aspect here is that realization of a definite orientation of a specific pair of polyhedrons automatically leads to the outcome when subsequent polyhedrons would be attached to that "leading" associate in the same way that causes the appearance of ordered macro clusters. Such system has a definite structure, which can explain the possibility of global structuring of large volumes of water. This aspect is adequately examined in the work of Zenin (1999).

The other hierarchical level (micro level) of the water structure is related to processes of movement and distribution of separate H_2O molecules between micro cavities of the spatial clathrate water frame and quasi-amorphous non-structured water. That micro level determines non-stationary evolution of H_2O molecules. The process of evolution is determined by two possible directions: molecules can leave the volume of quasi-amorphous water, penetrate the volume of these micro cavities and stay there for a long time in hydrophobic form, or, to the opposite, transfer from micro cavities into the volume of quasi-amorphous water.

It is absolutely clear that the micro level of water structure is distinguished by a much greater stability with respect to effects of external destructive factors than the macro level. With all external transformations of the clathrate frame typical for the macro level, hydrophobic H_2O molecules remain in a stable state in the volume of micro cavities. Such stability makes the micro level of water structure an effective object for organization of a system of water memory. Such system of memory has not been examined by anyone before.

We shall demonstrate how the presence of a clathrate frame of water may lead to formation of long-term memory in it and to recording and use of information.

Let's examine initial water in the state of thermodynamic balance with a certain temperature T. This condition is distinguished by maximal entropy. Such water is obtained by long boiling and slow cooling down or by a very long assertion. In this case the number of micro cavities within the system of clathrate hydrates filled with water matches the Bolzman distribution accounting for statistical weights of H_2O molecules' condition in micro cavities and in amorphous water. It would be balanced or regular water.

At the temperature of 4°C 18% of all micro cavities are filled with water, at the normal body temperature (36.6°C) 38% of micro cavities are filled, while at 55°C about 50% of micro cavities will be occupied by H_2O molecules.

This pattern of distribution is related to several aspects:

1. The Bolzman distribution at a given temperature;
2. The repetition factor of degeneration of the initial and final state of an H_2O molecule in amorphous water near an entry window into the volume of a micro cavity and inside it;
3. The ratio of the volume of all amorphous water and the volume of a clathrate frame.

With changing temperature all three values also change, which makes exact calculations of the dynamics of occupation of micro cavities more difficult. However, it is obvious that in the volume of clathrate micro cavities the energy of molecular links is close to zero (due to hydrophobic nature of interaction with the walls), while condition of an H_2O molecule in the volume of quasi-amorphous water is determined by the depth of a potential pit, dictated by links with other water molecules. The depth of that pit corresponds to the energy of activation at diffusion $\Delta W \approx 0.2 \ eV$, which, in effect, reduces the level of energy of an H_2O molecule with respect to condition of the same molecule in the clathrate frame by the value of ΔW. For these reasons, it becomes clear that a necessary energy of activation for entering a micro cavity ΔE_M and exiting from it $\Delta E_M - \Delta E$ would be different (*Fig. 2.1*). According to this the time of staying of an "extra" molecule of water in a micro cavity and the time of existence of an "extra" vacancy in an empty micro cavity would also be different.

With violation of thermodynamic balance a redistribution of H_2O molecules between amorphous water and micro cavities takes place until a new balanced state is achieved. We shall demonstrate that a spontaneous transfer between these states is substantially inhibited due to a very small probability of tunnel penetration of H_2O molecules through "narrow" windows and the time of existence of each of these conditions turns out to be very big. We shall determiner the time of relaxation in such redistribution.

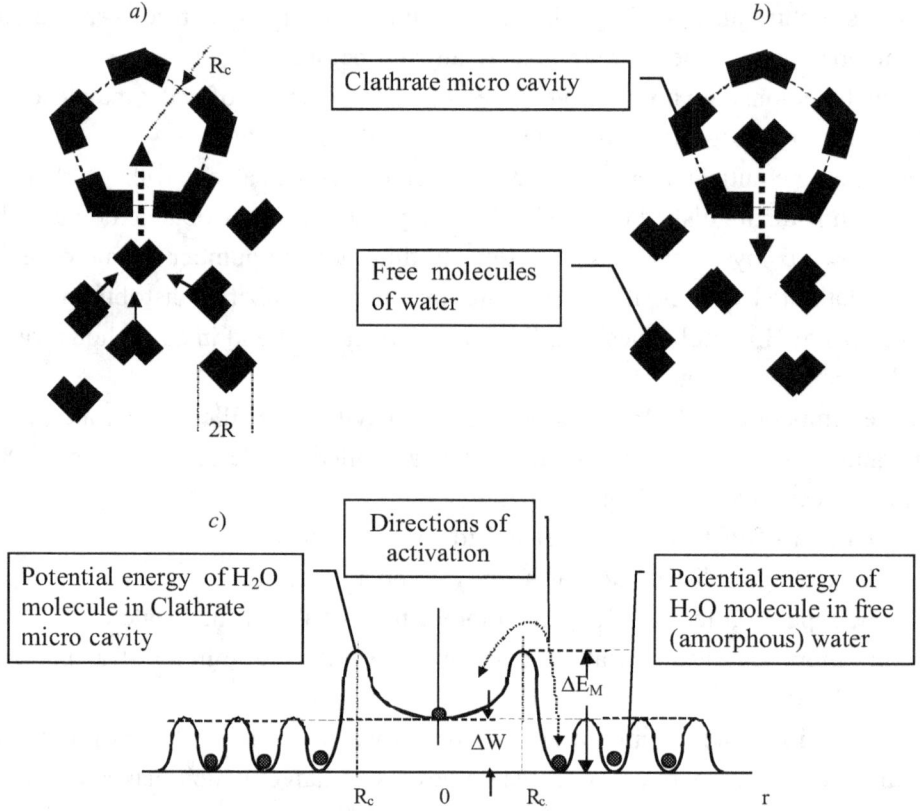

Fig. 2.2. Process of thermally stimulated activation a) and deactivation b) of micro cavities of a clathrate water frame at increasing and decreasing temperatures; c) structure of potential energy of molecules of amorphous and linked water in the volume of a clathrate micro cavity and around its boundaries

Such relaxation corresponds to a transfer of water molecules in two possible directions: a) from the state of amorphous water into the volume of a micro cavity (if the initial number of water molecules in micro cavities was smaller than the value of a variable determined by the Bolzman distribution, which may happen in case of a rapid heating of water); b) from the state of "excess" water in micro cavities to amorphous water (if the number of water molecules in micro cavities exceeded the value for balanced state, which, for example, corresponds to the case of a rapid cooling down of water).

The process of relaxation of each of these states depends on thermodynamic probability

$$W = exp(-\Delta E_M / k_B T) \qquad (2.2)$$

that one of the water molecules after its interaction with other molecules will receive energy ΔE_M sufficient for such its short-term deformation (i.e. work on increasing energy of interaction between a proton and an ion of oxygen), which would be enough for reduction of a water molecule's size to dimensions of a window of a micro cavity and, correspondingly, penetration of that molecule inside the micro cavity.

Since the frequency of collisions of any water molecule with the surface of any structured object in water is equal to the frequency of vibrations of molecules around a local position of balance $\omega_D \approx 1/\tau_0 \approx 10^{13}$ s, total probability of capturing a molecule in a unit of time into an empty micro cavity is equal to $F = W/\tau_0$. From this expression we can determine average lifetime of an unbalanced (empty) state of a single micro cavity in the volume of a spatial tetrahedral water frame (time of relaxation of a vacant spot in a micro cavity)

$$T_{1W} = 1/F_1 = \tau_0 \; exp(\Delta E_M/k_B T) \qquad (2.3)$$

Plainly, that time will determine the duration of water memory at filling up of that micro cavity (for example, during heating of water).

It is possible to calculate the value of ΔE_M, characterizing that process.

Fluctuational movement of a proton in a molecule H_2O in a direction, perpendicular to the line of link OH constitutes that of a harmonic oscillator (*Fig. 2.3*).

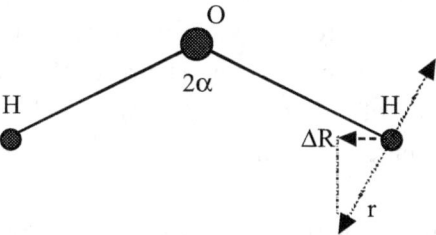

Fig. 2.3 Normal vibrations of a proton in a molecule of water

The potential energy representing shifting of an ion of hydrogen by the value r with respect to the point of balance may be expressed in the form of energy of a harmonic oscillator

$$V(r) = M_H \omega_H^2 r^2/2 \qquad (2.4)$$

Here M_H – mass of an atom of hydrogen, $\omega_H \approx 3*10^{14} \; c^{-1}$ – frequency of normal vibrations of a proton in a molecule of water in a direction perpendicular to the line of link OH (Zatsepina, 1999).

Taking into account that the angle between the lines of linking of each one of the protons with the nucleus of oxygen equals $2\alpha \approx 104.5°$, we find that energy required for deformation of the outer dimensions of a water molecule by the value of $\Delta R \approx 0.26 \; A$, sufficient for putting a molecule inside a micro cavity would be

$$V(\Delta R) = M_H \omega_H^2 \Delta R^2/2cos^2\alpha \approx 1.1 \; eV \qquad (2.5)$$

This variable represents energy threshold $\Delta E_M = V(\Delta R)$ determining the process of water relaxation. This threshold greatly exceeds thermal energy of water molecules equal to

$k_B T \approx 0.025 \ eV$ at room temperature. It can be seen, that time of relaxation T_{1W} is strongly correlated with the threshold value for energy of deformation of a water molecule ΔE_M and its temperature T. A big value of ΔE_M leads to a small probability of overcoming the barrier in the area of an entrance window to a micro cavity. In the result, the probability of spontaneous deactivation of water is very small, which means a very long period of storing information.

Let's make some quantitative estimates. At the temperature of water $T = 293K$ (20°C) the time of relaxation (duration of "water memory") is equal to $T_{1W} \approx 10$ days. With higher water temperatures time of relaxation decreases sharply and increases with its cooling down (see *Table 2.1*).

For an alternative direction of relaxation (transferring a single molecule of water from the volume of a micro cavity into the volume of amorphous water) time of relaxation T_{2W} is also determined by an expression similar to (2.3), where energy of activation is different ($\Delta E_M - \Delta E \approx 0.9 \ eV$ instead of $\Delta E_M \approx 1.1 \ eV$). Moreover, it is necessary to remember that because the inner dimension of a micro cavity is considerably bigger than the size of a potential pit for each molecule in the volume of quasi-amorphous water, actual frequency of collisions of a water molecule with the walls inside a micro cavity $\omega_D \approx 1/\tau_0$ will be lower, while period τ_0 will be, respectively, larger than in the volume of water. The results of calculation of time of relaxation T_{2W} during a reverse transferring of molecules H_2O from a volume of a micro cavity into quasi-amorphous water are also presented in *Table 2.1*.

Table 2.1 Relation of time of relaxation of water (duration of "water memory") to its temperature

$T^o \ C$	1	10	20	30	36.6	40	50	60	70	90
T_{1W}	300 days	49 days	10 days	58 hours	24 hours	15 hours	4.4 hours	1.3 hours	27 hours	3 hours
T_{2W}	30 min	14 min	4 min	1.5 min	45 sec	30 sec	12 sec	4 sec	1.5 sec	0.3 sec

It should be noted that in order to calculate values of T_{1W} and T_{2W} we need to know exact values for energy of activation and height of a potential barrier regulating entrance into volume of micro cavities of the clathrate frame. These parameters have been determined by us from model calculations. Respective values of T_{1W} and T_{2W} may differ considerably in specification of these parameters.

Obtained values of T_{1W} and T_{2W} represent water relaxation from an unstable state to a stable one, corresponding to a specific temperature.

If we examine this process from the point of view of information theory, generation of an unbalanced distribution may be considered a process of recording information in a volume of water. We shall call such water activated.

A very long time of relaxation T_{1W} allows to assume that water is a two-level (or rather double-zone) bi-stable system with a long lifetime in each of those states. Such system allows recording and holding of information (in the form of ratio of occupied and vacant micro cavities) as well as using that information effectively thanks to altering properties of water at time of transferring of a large number of H_2O molecules and other atoms dissolved in water, as well as molecules and ions from the state of amorphous water into the volume of linked micro cavities or the other way around (*Fig. 2.2*).

Although time of reverse relaxation T_{2W} at the exit of water molecules from the volume of micro cavities turns out to be considerably lower than time of direct relaxation at the entrance to these micro cavities, in any case it is by many orders greater than typical time of relaxation (2.1) $\langle\tau\rangle \approx 10^{-9} - 10^{-10}$ *s* due to fluctuations of the hydrogen link in the volume of amorphous water.

It may also be noted that the process of activation of water may be conducted not only during its heating or cooling, but also through effect of magnetic fields or ultrasound. Such periodic coherent influences can stimulate formation of quasi-stable clusters, each one of which uniting several reciprocally arranged cluster frames. In such a system behavior of isolated water molecules in periodically situated micro cavities is similar to movement of hydrogen in palladium, where a very high saturation of lattice is achieved. Periodic influences can also affect the parameters of the clathrate frame of water altering, for example, transparency of the potential barrier in windows of micro cavities (this is the problem about tunneling a molecule H_2O through a non-stationary barrier). Moreover, a strong periodic magnetic field can stimulate transfers between energy levels, which characterize the state of H_2O molecules in micro cavities and amorphous water (for example, due to multi-photon non-linear processes during interaction with magnetic moments), which causes uneven populating of micro cavities by water molecules and constitutes activation of water.

Why will there be a process of "degradation" of spatial tetrahedral water frames under impact of an alternating magnetic field? It is easy to verify that in many cases it would be an energy efficient process.

Apparently, at a low frequency of an external magnetic field (including an external magnetostaticfield) water is a diamagnetic characterized by negative static receptivity. At 20°C molecular magnetic receptivity of water is equal to $\chi = -1.297*10^{-5}$. Because $\chi < 0$ magnetic permittivity of water will be $\mu = 1 + 4\pi\chi < 1$. For comparison, magnetic permittivity of water is lower than that of vacuum ($\mu = 1$) and air ($\mu \approx 1$).

Density of energy of a magnetic field with strength $H(r)$ and induction $\vec{B}(r) = \mu\vec{H}(r)$ in a material environment with constant magnetic permittivity m is determined from the expression

$$u_M = \vec{B}(r)\vec{H}(r)/8\pi = \mu[H(r)]^2/8\pi \qquad (2.6)$$

Correspondingly, full energy of a magnetic field in a part of environment limited by volume V, is equal

$$u_M V = \mu V[H(r)]^2/8\pi \qquad (2.7)$$

Volume density of forces acting on a unit of volume of water equals

$$\vec{f}_M = (\mu - 1)\nabla \vec{H}^2 / 8\pi = -(1 - \mu)\nabla \vec{H}^2 / 8\pi$$

Apparently, in the case of diamagnetic (at $\mu < 1$) force of magnetic pressure will be directed against the gradient of magnetic field and will tend to change the location or shape of the sample in such a way so to move part of it to the area of a weak magnetic field.

Exactly the same situation occurs in water affected by a limited in space alternating magnetic field. At the exit of a molecule from a cavity the volume of water will increase, which will correspond to transferring of part of the full volume of water into an area with a weaker field.

With the help of such external influence an unbalanced population of micro cavities may be achieved, which would be unattainable through changing temperature. Similar activation of water may also be achieved by putting it under pressure from all sides.

2.3   SOME PHYSICAL AND MOLECULAR FEATURES OF ACTIVATED WATER

The issue of comprehensive study of features of activated water has not been addressed properly. Normally, only isolated characteristics (for example, viscosity, electric conductivity, the pH value) are examined and only under specific, often random, conditions. First of all, such water will have other physical and chemical characteristics. It should have a very specific effect on living objects.

High thermal stability of a human body automatically leads to a situation, when all water contained in it must have a fixed number of occupied micro cavities corresponding to normal temperature (*Table 2.1*). If activated water penetrates the body it causes alteration of parameters of the water environment in it. It can be observed from the results of calculations that at normal body temperature activated state of water achieved by a certain way may be preserved for 24 hours, which would be quite enough for its therapeutic effect.

If, for example, water was obtained by rapid heating, it will have an excess of both amorphous water and vacant micro cavities in the volume of the clathrate frame. Such water will have a lower volume density, which may significantly lesser burden on the heart and other human organs. Its viscosity will be much lower, which may substantially improve delivery of salts in the organism. Changing viscosity of water also affects the process of non-enzymatic self-reparation of double radiation breaks of DNA (Pinchuk, 2001; Vysotskii, 2002). A significant change of dielectric permittivity of activated water in the *UV* range of the specter produces the same result (Pinchuk, 2001; Vysotskii, 2002).

In activated water obtained, for example, by rapid cooling there would be a deficit of amorphous water and an excess of occupied micro cavities. The same constitution will be characteristic of water subjected to a long powerful compression and water from mountain springs (water molecules will be "squeezed" inside micro cavities and will not have enough time to leave them after compression is stopped). Such water has higher density and viscosity. During its relaxation there will be exiting of isolated H_2O molecules from micro cavities into the volume of amorphous water. These molecules can neutralize free radicals with its free links.

The obtained results provide a plausible explanation to a well-known method in medical practice – in order to keep water in its initial heated condition in activated state it is necessary to cool it down as fast as possible to a low temperature (for example, to a temperature of a living organism). This result follows immediately from the provided analysis of processes of direct and reversed relaxation of the process of preserving water memory based on different times of relaxation after establishing thermodynamic balance in two connected systems – the system of clathrate micro cavities and the system of quasi amorphous water. In case of an excess concentration of water molecules in the volume of clathrate frame, which is the case during cooling of water, the process of relaxation and, respectively, loosing of information occurs much faster that during its heating, when there is an excess of vacant micro cavities in the same clathrate frame. These results are presented in *Table 2.1* and refer to various values for energy of activation and deactivation of population in micro cavities.

Clearly, medical aspects of effects of two types of activated water deserve thorough clinical testing.

Based on such a system of long-term memory interpretation can be given to many effects related to water activation and manifestation of its anomalous properties. In conclusion, we shall note that the examined phenomena pertain to clear water and do not account for the influence of dissolved impurities (including ions and micro particles of iron), the presence of which may cause other effects (for example, at the presence of an external constant magnetic field).

An indirect confirmation of the premise that the process of activation may be related to changing of occupancy of micro cavities in the clathrate frame and transferring of water molecules from a linked state in the volume of micro cavities to a state of amorphous water is the process of spontaneous and fading luminescence of water. Several researchers observed that effect on many occasions. The reason for such luminescence may be found in local emission of energy of activation of a single molecule after crossing of a potential barrier regulating entry and exit from a micro cavity in the clathrate frame. This excess energy may be detected either immediately upon completion of crossing the barrier or through its catalytic effect of several physical and chemical transformation causing luminescence.

For example, Dr. Voeikov from the Moscow State University has observed the effect of luminescence by using the methods of chemiluminescence analysis after adding salts of double-valence iron and luminol as a fluorophor to water. Specifically, such fading luminescence can be observed in artesian water as well as water, which was preliminary stored in a closed bottle for a long period of time. It is interesting to note that temporal dependency of luminescence intensity was, in turn, closely dependent on material of bottle (glass, ceramics or plastic). Duration of luminescence at room temperature was 5–7 days, which well corroborates the data presented in *Table 2.1*. In settled balanced water as well as water preheated to a high temperature luminescence is not observed.

Literature to Chapter 2

Bernal J.D., Fowler R.H. // J. Chem. Phys. v. 1 (1933) *p. 513*

Davenas E. et al, // Nature, 1988, v. 333, *p. 816-818,* 1988

Klassen V.I. Water and Magnet, Moscow, Science, 1973

Nemethy G., Seheraga H.A. // J.Chem. Phys. v. 36 (1962) *p. 3382*

Pauling E. Hydrogen bonding // Ed. D. Hadzi, London, Pergamon Press, 1959

Pinchuk A.O., Vysotskii V.I. // Physical Rev E, v. 63 (2001) *p. 31904*

Samoilov O.Ya. // Structure of water solutions and hydration of ions, Moscow, 1957

Vysotskii V.I., Pinchuk A.O., Kornilova A.A., Samoylenko I.I. // Radiation Physics and Chemistry, v. 65 (2002), *p. 487*

Zenin S.V. Structured state of water as a basis of control of behavior and safety of living systems // MD Dissertation (Moscow Institute of traditional methods of treatment), Moscow, 1999

Zatsepina G.P. Physical properties and structure of water, Moscow State Univ. Publ. House, 1998

3. PATTERNS OF INTERACTION OF BIOLOGICAL MACRO MOLECULES IN ACTIVATED WATER

3.1 THE ISSUE OF RADIOACTIVE STABILITY AND PATTERNS OF INTERACTION OF BIOLOGICAL MOLECULES IN WATER ENVIRONMENT IN LIVE ORGANISMS UNDER EXTERNAL INFLUENCE

3.1.1 General patterns of long-distance dispersion interaction of biological molecules and their assemblies in a water environment

It is well known that water (in the form of a water-salt solution) is the foundation of the molecular-structural organization of all vital parts of any biological system. All inner cellular processes (including processes related to a cell's reaction on radiation and other types of influences) occur in a water environment and their efficiency is largely determined by water. The importance of water's influence on life support processes may be related to the effects of "water memory", which were discussed in chapter 2. Since activated water has several peculiar properties (it also concerns the dispersion and electrodynamic parameters of water), it is obvious, that the influence of activated water can have both positive and negative effect. It depends on a particular situation.

Without going too deep into discussion of mechanisms, by which water affects these processes (it will be done in further sections) note, that interaction of biological molecules in a water-salt environment is one of the main instruments of maintaining an organism's protection against effects of ionizing and non-ionizing radiation, as well as influence of free radicals.

Let's investigate this problem more closely. All systems of a live organism are subject to constant external and internal influence from many physical and chemical factors of natural and man-caused origin. There are many different negative consequences of such influence.

One of the effects of such influence, along with other lesser effects, may be a structural disturbance of a biological macromolecule. Such disturbances affect the whole hierarchical system of an organism. The most dangerous effect of an external influence is generation of double breaks of DNA macromolecules responsible for replication and transcription of new DNA macromolecules, which can, therefore, lead to a loss or distortion of genetic information stored in DNA (Kantor, 1995).

The most dangerous is the effect on DNA macromolecules free of protein, which contain the genes responsible for self-reproduction of DNA and the genes controlling the synthesis of proteins-enzymes. Due to absence of a reflecting protein shell, this most functional group of DNA macromolecules is particularly sensitive to radiation by a portion of fissionable (inter phase) and maintaining biosynthesis (differentiated) cells. Any organism has about 4% of DNA macromolecules free of protein shell from the total number of DNA macromolecules, but they constitute the sensitive element, which determines sustainability of the whole biological system.

Biological consequences of such disturbances are manifested in mutations and extermination of cells. It is necessary to note, that each cell has mechanisms of enzymatic self-reparation of

DNA partially offsetting the negative influence of mutation and degradation. For example, single breaks of DNA are eliminated with the help of a *ligase*. However, all, without an exception, enzymatic mechanisms are incapable of eliminating double breaks of DNA.

Obviously, the only possibility of solving the problem of DNA self-reparation (self-liquidation of double breaks) may be related to the long-distance electromagnetic interaction between either certain single nucleotides or their oriented pairs (Vysotskii, 1997).

Such interaction depends on the type of nucleotides, parameters of inner cellular media and width of a double break of the longitudinal spiral of a macromolecule R.

This section provides a detailed analysis of all features of total long-distance intermolecular interaction of end pairs of nucleotides in an inner cellular water-salt environment, which are located on the ends of a severed double-string DNA spiral. Beside that, some general questions of such interaction's dependency on changing parameters of an inner-cellular water-salt environment and characteristics of external influence on a biological system are addressed. An account of all features of an electrostatic interaction of ion charges distributed according to the surface structure of opposing nucleotides at the presence of an inner-cellular water-salt environment will be provided in the next section, together with numerical calculations of the structure of a potential pit.

An analysis of the efficiency of all possible forces acting in a space filled with water, between nucleotides located on the ends of a ruptured DNA spiral, as well as between all parts of lengthy DNA strings with periodically placed foundations shows that two equal forces have significant influence on spatial location of DNA fragments:

a)   long-distance fluctuational induction-dispersion force of Van Der Waals type for intermolecular interaction;

b)   Coulomb interaction of charges distributed on the surface of separate nucleotides providing for the influence of a water-salt environment in the space between them.

We know, that Van Der Waals force is of electric nature, although it is not related to electrostatic interaction (London, 1930; Barash, 1988). It is normally regarded as a so-called dispersion or electrodynamic force. Let's examine some features of the Van Der Waals force.

Each of the molecules or atoms may be characterized by a dipole moment. Dipole moment expresses asymmetrically placed and distributed electric charges in an electrically neutral system. Dipole moment of a small molecule matches, by the order of its size, the product of an electron's charge by the length of a chemical link. For processes of molecule interaction induced dipole moments emerging due to influence of an external field, have special significance.

Appearance of an induced dipole moment is defined by polarizing ability of an object. Polarizing ability characterizes ability of an electronic shell of the same molecule or an atomic group to become deformed and shift under the influence of an external electric field $\vec{E}(t)$. This field induces an electric dipole moment $\vec{p}(t) = \vec{\alpha}\vec{E}(t)$ in an atomic or molecular system.

The value $\vec{\alpha}$ measured in terms of volume represents polarizing ability. In the general case, the polarizing ability is a tensor variable.

If another dipole $\vec{p}_1(t)$ is the source of the field $\vec{E}(t)$, such induced mechanism allows for interaction between dipoles. This interaction would also take place if generation of the initial dipole $\vec{p}_1(t)$ is purely accidental or fluctuational. The reason for that could be quantum fluctuations of a charge distribution in an atom or molecule. An accidentally generated dipole forms a random fluctuating field $\vec{E}(t)$ in the close surrounding space, which, in its turn, causes appearance of an induced dipole moment $\vec{p}_1(t)$ in another atom or a molecule and ultimately results in interaction of dipoles with energy V, which is proportionate to the product of dipole moments, i.e. $V \sim \vec{p}_1(t)\vec{p}_1(t)$.

Treating two molecules in the main "s-state" at a distance R from each other like two harmonic oscillators with frequencies ω_1 and ω_2 and polarizing abilities α_1 and α_2, their energy of interaction would be determined by the formula (London, 1930)

$$V(r) = -3\eta\omega_1\omega_2\alpha_1\alpha_2/2(\omega_1 + \omega_2)R^6 \qquad (3.1)$$

If we take atomic polarizing ability of an oscillator without taking account of its dispersion in the form

$$\alpha_i = e^2/m\omega_i^2 \qquad (3.2)$$

interaction of identical molecules assumes the form

$$V(r) = -3\eta\varepsilon^4/4m^2\omega_1^3R^6 \qquad (3.3)$$

Here, m – mass of an electron and e – its charge.

Precise calculations based on equations of quantum mechanics give the expression for the energy of interaction of neutral atoms or molecules

$$V(r) = -\eta e^4/8m^2\omega_1^3R^6,$$

different from modeled results (London, 1930) only by a numeric coefficient.

Thus, the modeling problem of interaction of two specific atoms at a long distance without taking dispersion of their polarizing ability into consideration can be easily solved.

Stating the problem about interaction within biological systems would, naturally, be absolutely different – there is a large number of atoms and molecules with strong electromagnetic dispersion and mutual reflection of each other in both parts of the ruptured DNA spiral. Beside that, it's contained in a water solution with strong frequency dispersion of dielectric permittivity $\varepsilon(\omega)$.

The fundamental thermodynamic reason for appearance of Van Der Waals forces is change of energy in a system of oscillators (atoms, molecules) during their rapprochement. The nature of appearing of these forces lies in quantum electrodynamics. In the contemporary interpretation of quantum electrodynamics, an electromagnetic field is regarded as a combination of mutually independent electromagnetic modes, each of which being a kind of a harmonic oscillator, where there is continuous transformations of the electric and magnetic components of a field. Like in any oscillator, the minimum value of energy of a separate mode corresponds to zero fluctuations, is different from zero and also depends on frequency of the mode. The reason for appearance of these zero fluctuations is linked to the ratio of uncertainties. Total energy of all modes of a field depends on the number and structure of the modes. With any change of the spatial configuration of a given volume in space, the structure of a fluctuating electromagnetic field in these areas and its total energy change. The number of these modes

$$N = \int \rho(\omega_n) d\omega_n,$$

their spectral density $\rho(\omega_n)$ and frequency position of separate modes of field ω_n is determined from border conditions of the surface of a selected volume of space.

Therefore, full volume density of energy of a field depends on properties of interacting objects. On the other hand, the derivative from total energy by volume (with constant entropy) determines the sign and the value of density on surfaces of interacting objects.

Clearly, we see that in order to change that energy certain work should be done with application of certain force. At the same time, if a system's energy falls, this work is done by the field, which corresponds to mutual attraction of the walls limiting a selected volume of space.

In the opposite case, when energy of a field increases, this work should be done by external forces and this corresponds to mutual repulsion of the walls.

We shall give some results.

In the case of interaction between unlimited mutually parallel metal surfaces with distance L between them, the force of pressure per unit of area equals (Dzyaloshinski, 1961; Barash, 1988)

$$P = \pi^2 \eta\_/240L^2 \qquad\qquad (3.4)$$

This interaction always corresponds to attraction. With a change to limited by their cross dimensions interacting surfaces, this force decreases, while with a certain ratio of dimensions and shape it can even change its sign. For example, two convex metal half spheres do not attract, but rather pushed apart (Dzyaloshinski, 1961).

We conclude from this simple example, that a complex change of the field structure in the space between interacting objects may lead to ambiguous sign for forces of interaction: objects of one shape attract, while objects of another shape repulse. As we shall demonstrate later, same features relate to different materials, from which interacting objects are made. Using such special features of interaction allows creation of systems of mutual recognition, making selection depending on the shape of a surface and material from which such surface is made. Obviously, such mutual recognition can play a very important role in biological systems and, particularly, in the immune system of an organism.

Accounting for frequency-dispersion structure of a substance of interacting objects (their dielectric permittivity $\varepsilon_1(\omega)$ and $\varepsilon_2(\omega)$) as well as environment in the space between them (it's characterized by dielectric permittivity $\varepsilon_w(\omega)$) makes analysis much more complicated.

The simplest case is interaction of two flat objects placed at a distance $L > \lambda_0$. The border wave length $\lambda_0 = 2\pi c/\omega_0 \, \varepsilon_1(\omega_0)$ corresponds to an average frequency ω_0 of the group of most powerful electromagnetic resonance in these objects. Such limitation results in a situation when in making calculations of energy of interaction it is sufficient to take into account only that area of frequencies ω, which satisfies the condition

$$\omega < \omega_0 < \omega_L \equiv 2\pi c/L$$

This limitation corresponds to "low frequency approximation".

In this case, pressure on the surface of interacting objects is determined by the expression

$$P = (k_B T/8\pi L^3)(\varepsilon_{10} - \varepsilon_{W0})(\varepsilon_{20} - \varepsilon_{W0})/(\varepsilon_{10} + \varepsilon_{W0})(\varepsilon_{20} + \varepsilon_{W0}) \qquad (3.5)$$

Here, all values for permittivity correspond to small frequencies:

$$\varepsilon_{10} = \varepsilon_1(\omega < \omega_L), \ \varepsilon_{20} = \varepsilon_2(\omega < \omega_L), \ \varepsilon_{W0} = \varepsilon_W(\omega < \omega_L)$$

Since the strongest resonance in atomic systems belong to the ultra violet frequency range, normally, the border wave length $\lambda_0 \approx 0.1$–0.05 microns.

For small distances $L \ll \lambda_0$ the force of pressure is determined by a much more complex function (Dzyaloshinski, 1961)

$$P = (\eta/8\pi^2 L^3) \int S(i\xi)d\xi \qquad (3.6)$$

In this expression the value

$$S(i\xi) = [\varepsilon_1(i\xi) - \varepsilon_W(i\xi)][\varepsilon_2(i\xi) - \varepsilon_W(i\xi)]/[\varepsilon_1(i\xi) + \varepsilon_W(i\xi)][\varepsilon_2(i\xi) + \varepsilon_W(i\xi)]$$

depends on frequency dispersion of dielectric permittivity $\varepsilon_i(i\xi)$ of all environments located in the area of interaction, provided that an "imaginary frequency" $\omega = i\xi$ should be used as an independent variable for dielectric permittivity. The need for making this substitution is dictated by the principle of causation and emerges in case of using complex integration.

From general formulae (3.5) and (3.6) we can see that during interaction of identical objects, which corresponds to equality of values for dielectric permittivity $\varepsilon_1(\omega) = \varepsilon_2(\omega)$, direction of pressure and the sign of the force applied to interacting objects are always positive. This result corresponds to mutual attraction of objects irrespective of characteristics of the force, which separates these objects.

During interaction of different objects, direction of pressure depends on the ratio of $\varepsilon_1(\omega)$ and $\varepsilon_2(\omega)$.

Specifically, if $\varepsilon_1(\omega) > \varepsilon_W(\omega)$, $\varepsilon_2(\omega) < \varepsilon_W(\omega)$ or if $\varepsilon_1(\omega) < \varepsilon_W(\omega)$, $\varepsilon_2(\omega) > \varepsilon_W(\omega)$ over the whole range of frequencies, the sign of the variable for pressure will be negative, which corresponds to mutual repulsion of objects.

If, however, $\varepsilon_1(\omega) > \varepsilon_W(\omega)$, $\varepsilon_2(\omega) > \varepsilon_W(\omega)$ or $\varepsilon_1(\omega) < \varepsilon_W(\omega)$, $\varepsilon_2(\omega) < \varepsilon_W(\omega)$ objects will always attract to each other.

In the general case, the sign of the force of interaction between objects and the direction of corresponding pressure on the surface of these objects depend on characteristics of dielectric permittivity in the entire range of frequencies. There is a typical situation, when in one part of the specter $\varepsilon_1(\omega) > \varepsilon_W(\omega)$, $\varepsilon_2(\omega) > \varepsilon_W(\omega)$, while in the other $\varepsilon_1(\omega) > \varepsilon_W(\omega)$, $\varepsilon_2(\omega) < \varepsilon_W(\omega)$ or $\varepsilon_1(\omega) < \varepsilon_W(\omega)$, $\varepsilon_2(\omega) > \varepsilon_W(\omega)$. Such patterns lead to the situation when the value $S(i\omega)$ becomes a function of frequency with variable polarity. Because of that, interaction corresponds to attraction of objects in one area of frequencies, while producing repulsion in the other one. A final force is determined by an algebraic sum of all variable-polarity components from different sectors of the electromagnetic specter.

Therefore, we conclude, that a controlled change of dispersion characteristics of either each of the objects or the environment between interacting objects makes it possible to affect the sign and the value of the force of interaction between objects. Comparing this conclusion with the problem of stability of biological systems, we arrive at a conclusion that because we cannot change characteristics of interacting biological molecules, the only possibility of managing their interaction and mutual recognition is precisely related to influencing water, which is the foundation of the inner cellular liquid. There are many ways of affecting water and the process of its activation discussed in chapter 2 is one of the most promising.

In case of a limited size of interacting objects with volumes V_1 and V_2, given the condition that the distance between these objects does not exceed their cross section dimensions (this condition corresponds, for example, to interaction between nucleotides of DNA, located on the ends of a ruptured DNA string), total force of interaction in the general case equals (Dzyaloshinski, 1961)

$$F(L) = (71\eta V_1 V_2 / 8\pi^3 L^7) \int_{\omega_{min}}^{\omega_{max}} S_1(i\xi)d\xi ; \qquad (3.7)$$

$$S(i\xi) = [\varepsilon_1(i\xi) - \varepsilon_W(i\xi)][\varepsilon_2(i\xi) - \varepsilon_W(i\xi)]/\{[\varepsilon_1(i\xi) + 2\varepsilon_W(i\xi)][\varepsilon_2(i\xi) + 2\varepsilon_W(i\xi)]\}$$

The DNA structure and positioning of nucleotides will be discussed in more detail later.

The upper limit of integration $\omega_{max} = c\delta/L$ in (3.7) is determined from the condition, that maximum frequency ω_{max} contributing to the force interaction (3.7) must be significantly lower than characteristic extreme frequency $\omega_0 = c/L$, determined by the distance between objects L.

This condition is related to the possibility of neglecting the effect of delayed fluctuational electromagnetic fields causing electrodynamic dispersion interaction at a small distance $L < c/\omega_0$.

The lower limit of integration ω_{min} in (3.7) is determined by the "cutoff" frequency. It is a frequency, below which spreading electromagnetic waves in any environment is impossible. Plasma-like environments (including electrolytes), containing free charges, are regarded as such material environments. Frequency range $0 < \omega < \omega_{min}$ is an interval with frequencies not participating in force intermolecular interaction.

Appearance of free charges in an environment can be fluctuational (for example, due to dissociation of atoms and molecules in an environment's thermal movement) as well as directed (in case of an ionizing impact on an environment).

Double breaks of DNA spirals appear due to impact of ionizing radiation or free radicals (Ryabchenko, 1979). If, for end nucleotides (i.e. for nucleotides located on both sides of a double break of DNA) condition $(\varepsilon_{10} - \varepsilon_{w0})(\varepsilon_{20} - \varepsilon_{w0}) < 0$ holds for large distances $L > \lambda_0$, the width of breaks L increases due to mutual repulsion of the ruptured ends of DNA. In the opposite case, when $(\varepsilon_{10} - \varepsilon_{w0})(\varepsilon_{20} - \varepsilon_{w0}) > 0$ with $L > \lambda_0$, there is mutual attraction of the ruptured fragments repairing a break. Note, that at a large distance, considering an unusually large value of static dielectric permittivity of water $\varepsilon_{30} \approx 80$ at a temperature, optimal for biostructures, there is probably mutual attraction. However, such super long distance interaction (in the scale of DNA dimensions) has no practical meaning, since it takes place at distances $L \approx 1$ micron, which by many orders exceed such values $L \approx 5-50$ A, which correspond to actual dimensions of DNA ruptures.

Interaction at small distances is much more important. It depends on the entire specter of frequencies of dielectric permittivity.

As the following quantitative analysis shows, a typical situation when the combination $[\varepsilon_1(i\xi) - \varepsilon_w(i\xi)][\varepsilon_2(i\xi) - \varepsilon_w(i\xi)]$ has different signs in different areas of the specter, which leads to mutual attraction of one nucleotides and mutual repulsion of others. Due to randomness of a break and random combination of end nucleotides it leads to a final impossibility of self-reparation of damaged DNA spirals.

Since dispersion parameters of the nucleotides are stable and they cannot be controlled, the only possibility of regulating the process of self-reparation of DNA is through controlling the dispersion parameters of water (including during its activation).

3.1.2 Dispersion electromagnetic characteristics of DNA

We shall conduct an analysis of dispersion characteristics of deoxyribonucleids. First, let us recall some necessary information about DNA structure (Ryabchenko, 1979, Zenger, 1997).

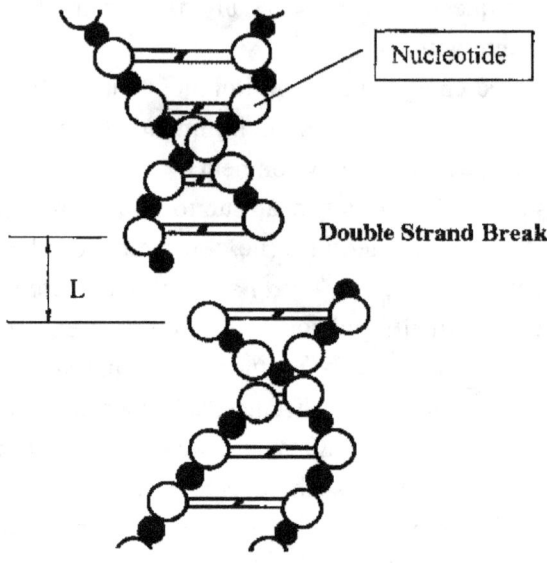

Fig. 3.1 Spatial structure of a DNA spiral in the area of a double break

Deoxyribonucleic acid (DNA) is a high molecular biopolimer with a linear chain consisting of alternating monomeric units – deoxyribonucleotides (shortly – nucleotides). Four types of monomers are normally found in natural DNA: deoxyadenosine monophosphate (adenin), deoxyguanozin monophosphate (guanin), deoxytimidin monophosphat (timin) and deoxycitidin monophosphate (citosin). Difficult names of these nucleotides reflect their complicated nature: each nucleotide consists of a carbohydrate (pentose) component, a purine or pirimidin nitrogen base and pyrophosphor acid residue.

Connection of deoxyribose and phosphate residues is a universal type of internucleotide link in DNA. Purine and pirimidin bases — adenine (*A*), guanine (*G*), thymine (*T*) and cytosine (*C*) – are attached to framework deoxyribosnophosphate chain (another frequently used term – sugar-phosphate chain) as side radicals.

In a double DNA spiral there is, on average, about the same quantity of four main nucleotides, provided that, naturally, the quantity of nucleotides included in their respective complimentary pairs *AT* and *GC* is matched exactly. The mole ratio of different pairs changes only slightly (by not more than 2–5%) in different organisms and also from molecule to molecule in same cells.

The basis of an informational structure of a DNA macromolecule is alteration of sequences of four basics – *A*, *G*, *T* and *C*. This sequence determines the code of hereditary information. DNA also includes – in very small quantities – analogues of the listed basics: oxyderivatives of adenin and guanin – hypoxantin (*HX*), xantin (*X*); demitilled timin – uracill (*U*); oximetilderivatives of uracill and timin; 5-metilcitosin.

Integrity of hereditary information during transformation is provided by a unique mechanism of fault-proof copying of a sequence of nucleotides by a newly synthesized molecule of DNA from already existing DNA as a matrix. The principle of complimentary character (i.e. mutual spatial compliance) being the basis of matrix mechanism of copying makes it practically fault-free. Statistical analysis shows that, on average, for one macro molecule of DNA containing no less than 10^7–10^8 pairs of nucleotides, there is only one incident of distorted information. As a rule, this distortion is related to the so-called tautomeric replacements. In the end, nucleotides constitute two complimentary semi-nucleotide chains, united by hydrogen links: pair *AT* by two and *GC* – by three.

Normally, DNA exists in the form of two complimentary chains forming a natural secondary structure – right double-string spiral (*Fig 3.1*). Nitrogen bases in that model are always directed inside the spiral (towards its long axis) and oriented perpendicular to the long axis with its plain. The bases of complimentary pairs are always located in the same plane.

Given a long enough chain, due to repulsion of its negative electric components, a rigid DNA configuration appears in a solution, characterized by a spiral period of 34 *A* and diameter of 20 *A*. In the process of forming such DNA spiral structure water surrounding a DNA spiral, plays a very important role. In the absence of water (for example, on the surface of covering glass) cross section of a molecule of DNA assumes the ellipsoid shape. The distance between adjacent pairs of nucleotides equals 3.4 *A* and each turn of the spiral contains 10 nucleotide pairs. A spiral with such parameters represents the so-called "stretched" *B*-form of DNA, which is an unstable meta stable form of a molecule, distinguished from the stable "compressed" *A*-form with eleven nucleotides by one full turn of the spiral and spiral step of 28 *A*.

A DNA spiral with step 34 *A* (*A*-form) corresponds well with the quasi-crystalline water structure discussed in chapters 1 and 2. DNA is capable of bringing water into order on the distance of up to 1000 *A* from its surface! Transfers between forms *A* and *B* in the result of dehydration of DNA is a non-specific reaction on very different forms of influence (radiation, heat, *pH*, cations) conditional upon breaks of hydrogen links, neutralization of phosphorus groups as well as suppression of their dissociation.

Twisting of the double-string DNA spiral into a third formation leads to either a disorderly ball or some true geometric shape (for example, an ellipsoid of rotation). A DNA structure of the third-formation in the form of regular secondary spiral is found in *T*-even protobes. A free protobic DNA in a protein case is rolled into a third formation spiral (four turns) with external diameter in the order of 232 *A* and internal cavity of about 80 *A* in diameter.

Parameters of the third-formation structure, a free DNA as well as parameters of the secondary formation structure significantly depend on the degree of molecular hydration. If the degree of hydration is high, a rigid secondary B-formation emerges, which rolls into a less compact "long spindle". If hydration is low, a "soft" secondary A-formation appears, which rolls into a more compact "short spindle". Specifically due to formation of the secondary and tertiary structures DNA performs the functions of chromosome formation and regulation of protein synthesis in a cell.

Regulation of protein synthesis and the process of self-reproduction is accomplished by a DNA free of protein casing, activated in the result of protein depravation and the $A{\rightarrow}B$ transfer.

Due to the fact that the character of interaction of fragments of biological molecules largely depends on their dispersion characteristics, we need to discuss the electromagnetic features of nucleotides.

The main resonance of absorption of DNA molecules and the infrared spectrum are in the range of $\lambda_{s1} \approx 6$–11 microns and $\lambda_{s2} \approx 3$ micron and explained by vibrations of the main molecular groups N-H, O-H and P-O-C (Vysotskii, 1997).

The presence of each of these resonance vibrations corresponds to a change of dielectric permittivity of nucleotides. In this process, the sum change from all resonance vibrations has the form

$$\Delta\varepsilon_{ir}(i\xi) = \sum_{s=1}^{2} \Delta\varepsilon_{s,ir}(0)/(1 + \xi^2/\omega_s^2); \qquad (3.8)$$

$$\Delta\varepsilon(0)_{1,ir} \approx 0{,}8; \; \Delta\varepsilon(0)_{2,ir} \approx 0{,}3$$

The next (in the order of increasing frequency) group of electromagnetic resonance vibrations of nucleotides lies in the near ultraviolet range of the spectrum. These resonance vibrations are related, first of all, to amide groups CO-NH and have the following parameters (see *Table 3.1*).

Individual value of contribution into total dielectric permittivity is determined by the expression

$$\Delta\varepsilon_{UV}(i\xi) = \sum_{L=1}^{2} \Delta\varepsilon_{L,UV}(0)/(1 + \xi^2/\omega_L^2) \qquad (3.9)$$

Table 3.1 Resonance dispersion features of water in the ultraviolet frequency range

Nucleotide	Resonance wave length λ_L, micron	Resonance frequency ω_L, $(10^{15}\ s^{-1})$	Amplitude of change of dielectric permittivity $\Delta\varepsilon_{L,UV}(0)$
Adenine	0.260 0.240	7.25 7.85	0.192 0.039
Guanine	0.278 0.251	6.7 7.5	0.072 0.198
Cytosine	0.271 0.240	7.0 7.8	0.114 0.096
Thymine	0.263 0.240	7.2 7.85	0.138 0.018

The last group of resonance vibrations, which ought to be considered in the analysis of the problem of intermolecular interaction involving dispersion features of certain DNA elements belongs to hard ultraviolet spectrum range (vacuum ultraviolet or *VUV*) and related to resonance of absorption in the vicinity of the first potential of ionization of atoms of carbon C (ionization potential is equal to $\eta\omega_C = 11.26\ eV$, $\lambda_C = 0.071$ micron), oxygen O and hydrogen H (respectively $\eta\omega_0 \approx \eta\omega_H \approx 13.6\ eV$, $\lambda_{O,H} \approx 0.084$ micron) as well as nitrogen N ($\eta\omega_N \approx 14.54\ eV$, $\lambda_N = 0.092$ micron) from the contents of thymine $C_4O_2H_4N_3$, cytosine $C_4OH_2N_3$, adenine $C_5H_3N_5$ and guanine $C_5H_2N_5$. In fact, frequencies of these resonance vibrations are determined by the K-edge of absorption known from Roentgen spectroscopy.

Table 3.2 Distribution of main chemical elements in the volume of nucleotides

	P_C	P_N	P_O	P_H
Adenine	0.38	0.38	–	0.24
Guanine	0.41	0.41	–	0.18
Cytosine	0.4	0.3	0.1	0.2
Thymine	0.3	0.23	0.16	0.31

Taking into account the weight (partial) share P_x of each of the listed atoms ($x = C, O, H, N$) in corresponding nucleotides (see *Table 3.2*) and proceeding from approximate equality of average volume concentration of atoms of water and nucleotides ($n_n \approx n_w$), the general structure of a corresponding element determining contribution of a given mechanism to the total dielectric permittivity has the form

$$\Delta\varepsilon_H(i\xi) = \sum_{x=1}^{4} \Delta\varepsilon_{Hx}(0)P_x/(1 + \xi^2/\omega_x^2) \qquad (3.10)$$

Here, $\Delta\varepsilon_{Hx}(0) = 4\pi n_n^2 e^2/m\omega_x^2$, provided that

$$\Delta\varepsilon_{HC}(0) \approx 0.17; \; \Delta\varepsilon_{HO}(0) \approx \Delta\varepsilon_{HH}(0) \approx 0.23; \; \Delta\varepsilon_{HN}(0) \approx 0.28$$

The final form of total dielectric permittivity of certain nucleotides contained in a DNA macromolecule (as a function of imaginary frequency) is characterized by a formula

$$\varepsilon_{1,2}(\xi) = 1 + \Delta\varepsilon_{ir}(i\xi) + \Delta\varepsilon_{UV}(i\xi) + \Delta\varepsilon_H(i\xi) \qquad (3.11)$$

and may be calculated with the help of the expressions provided above.

3.2   DISPERSION FEATURES OF THE DNA REPARATION SYSTEM

3.2.1   Kramerse-Kroninge method of reconstruction of dispersion characteristics of DNA nucleotides by experimental specters of absorption

The analysis of total dielectric permittivity of separate nucleotides in a DNA macromolecule conducted earlier is quite limited, not adequately precise and is normally useful only for qualitative analysis of dependency of the Van Der Waals interaction from dielectric features of a water-salt media and nucleotides. The reason for such situation is obvious.

Since the result interaction of separate nucleotides and their pairs is determined by the balance of forces of Coulomb interaction of charges, distributed on the surface of nucleotides, and forces of Van Der Waals interaction, having a different value and polarity in various intervals of distances, significant errors, inevitably linked to the use of modeled dielectric permittivity, unavoidably lead to very significant and unpredictable errors. Moreover, even small errors in determining the value of each of these forces may cause total indeterminacy in the value and polarity of the result interaction. Obviously, such contingency requires strict quantitative analysis for specific nucleotides. The most productive is the approach, according to which information about dielectric permittivity of nucleotides used in calculation of intermolecular forces of interaction between them is obtained from actual experiments aimed at determining spectral characteristics of optical absorption.

The idea behind such an analysis is based on several consecutive actions.

First, using Kramerse-Kroninge transformations (Landau, 1959), from initial experimental specters of absorption of nucleotides (represented by the coefficient of absorption $\alpha(\omega)$ or its equivalent – imaginary part $k(\omega)$ of the complex index of refraction $\mathring{n}(\omega) = n(\omega) + ik(\omega)$), reconstruction of the real part $n(\omega)$ of the complex index of refraction is performed for adenine, guanine, cytosine and thymine in infrared *(IR)* and ultraviolet *(UV)* ranges.

Connection between the imaginary part of the complex optical coefficient of refraction $k(\omega)$ and the real part of the same coefficient $n(\omega)$ is used in calculation. This connection is presented explicitly in the form of the Kramerse-Kroninge relation

$$n(\omega) = 1 + \frac{1}{\pi} P \int\limits_0^\infty \frac{k(x)}{x^2 - \omega^2} dx^2 \qquad (3.12)$$

In this expression, the symbolic operator P indicates, that integration is conducted for the main value by the Koshi method (Landau, 1959).

The imaginary part of the complex coefficient of refraction $k(\omega)$ is related to the optical coefficient of refraction $\alpha(\omega)$ of specific nucleotides with the help of the expression

$$k(\omega) = c\alpha(\omega)/4\pi\omega \qquad (3.13)$$

On the second stage, from the obtained complex index of refraction $n(\omega)$, the real $\varepsilon'(\omega)$ and the imaginary $\varepsilon''(\omega)$ parts of dielectric permittivity

$$\varepsilon(\omega) = \sqrt{\tilde{n}(\omega)} = \varepsilon'(\omega) + i\varepsilon''(\omega)$$

is calculated in corresponding ranges.

After that, we can find the explicit form of the relation of dielectric permittivity of nucleotides $\varepsilon(i\xi)$ and the imaginary frequency $\omega = i\xi$. As follows from the Lifshitz theory (Sabisky, 1973), and as we'll show later, the value of $\varepsilon(i\xi)$ is defining for calculation of forces of interaction between nucleotides.

Complex dielectric permittivity $\varepsilon(\omega)$ is related to the complex coefficient of refraction $\tilde{n}(\omega) = n(\omega) + ik(\omega)$ by dispersion relations

$$\varepsilon'(\omega) = n^2(\omega) - \kappa^2(\omega)$$
$$\varepsilon''(\omega) = 2n(\omega)\kappa(\omega) \qquad (3.14)$$

In order to find a relation of dielectric permittivity as a function of imaginary frequency, it is useful to use the equality, which follows from Kramerse-Kroninge relations (Landau, 1959)

$$\varepsilon(i\omega) = 1 + \frac{2}{\pi} \int\limits_0^\infty \frac{\varepsilon''(x)}{x} dx \qquad (3.15)$$

The need for introducing an imaginary frequency in dispersion relationships for dielectric permittivity of nucleotides is motivated (according to the theory of complex variable functions) by the lack of special points in the obtained dielectric function in the upper half-plain in the area of complex frequencies. Using a more conventional terminology, introducing an imaginary frequency is a necessary condition for fulfillment of the principle of causality in processes of electromagnetic interactions.

Accuracy of the obtained relationships for dielectric permittivity of nucleotides as a function of imaginary frequency depends on completeness of the experimental data for absorption of electromagnetic radiation on the entire spectral range, in other words, it depends on accounting for all existing resonance on the entire range of frequencies.

3.2.2 Numeric calculations of specters of the complex coefficient of refraction and dielectric permittivity of nucleotides in a DNA macromolecule

In performing quantitative analysis, known experimental data from reliable sources for specters of absorption of nucleotides in infrared and ultraviolet ranges was used (Cantor, 1984; Alberts, 1994). Beside that, specters of absorption in the infrared range, referred to in scientific projects, were also used (Kudryashev, 1982; Young, 1989). Intensive absorption of all nucleinic acids in the near *UV* range owes almost entirely to purine and pyrimidine bases. Sugar-phosphate framework makes an insignificant contribution to the specter of absorption at wave lengths, exceeding 200 *nm*.

We shall briefly consider the methodology of calculations (Pinchuk, 1999, 2001; Vysotskii, 2003) and provide the results.

Experimental specters of absorption for adenine, guanine, thymine and cytosine in water at *pH*7 in the ultraviolet range of the specter are shown in *Fig. 3.2a, 3.3a, 3.4a* and *3.5a* respectively. Note, that spectral bands of absorption of adenine and thymine with maximal wavelength λ_{max} = 260 *nm* have the form of solid (without fine structure) Gauss curves. In reality, these specters correspond to several very closely placed electronic passes. This problem was thoroughly discussed in the previous section.

Further, these experimental curves were extrapolated to the area of far ultraviolet specter and to the visible range. Then, using expression (3.12), calculation of the imaginary part of the complex coefficient of refraction $\kappa(\omega)$ of frequency was made, while the real part $n(\omega)$ of this complex coefficient was calculated on the basis of expression (3.14).

Once integration was done in predetermined ranges of frequencies, interpolation of the obtained discreet values of the real part of the complex coefficient of refraction and calculation of real and imaginary parts of the complex dielectric permittivity of the corresponding nucleotides were conducted.

On *Fig. 3.2b* there are dispersion relations of real and imaginary parts of the optical coefficient of refraction for adenine $n(\omega)$ and $k(\omega)$ made with the help of equations (3.12) and

(3.13). The real $\varepsilon'(\omega)$ and imaginary $\varepsilon''(\omega)$ parts of dielectric permittivity for adenine, calculated with the help of equation (3.14) are presented on *Fig. 3.2c*. Having obtained numerically the functional relationship of the imaginary part of dielectric permittivity for adenine in the ultraviolet range, relation between dielectric permittivity and the imaginary frequency in the upper half-plain was determined by simple integration of expression (3.15).

Functional dependence of the coefficient of absorption $\alpha(\omega)$ and dielectric permittivity $\varepsilon(i\xi)$ from imaginary frequency for adenine in ultraviolet range is presented on *Fig. 3.2d*. On the same figure there is dependence of dielectric permittivity from imaginary frequency $\varepsilon(i\xi)$ for adenine.

Similar dependence of the coefficient of absorption and dielectric permittivity of adenine in the infrared range, as well as function $\varepsilon(i\xi)$ is presented on *Fig. 3.2e*.

Expressions for the rest nucleotides are shown in *Fig. 3.3, 3.4* and *3.5*.

Conducted numerical calculations are based on the use of precise values for dielectric permittivity, obtained from actual experimental values for coefficients of absorption with the help of the causation principle, constituting the foundation of the Kramerse-Kroninge method of integral transformations. Advantages of that method are obvious. For example, from spectral dependency of these coefficients of absorption in infrared range (*Fig. 3.2e, 3.3e, 3.4e* and *3.5e*) it is easy to come to an obvious conclusion, that none of the model estimates can adequately represent such a complex and multifarious structure of these coefficients and, therefore, produce results of commensurable accuracy in further calculations.

The obtained dispersion dependencies for all major parameters ($n(\omega)$, $k(\omega)$, $\varepsilon'(\omega)$, $\varepsilon''(\omega)$, $\varepsilon(i\omega)$) for different nucleotides are the basis of calculation and analysis of forces of Van Der Waals interaction of end pairs of nucleotides on the whole range of distances from 3 to 20 A, which must be considered in the investigation of features of interaction of end pairs of nucleotides, located on different sides of a double DNA break. These parameters will be used in the next section for calculation of features of long-distance force interaction of end pairs of nucleotides in the area of double-string DNA breaks.

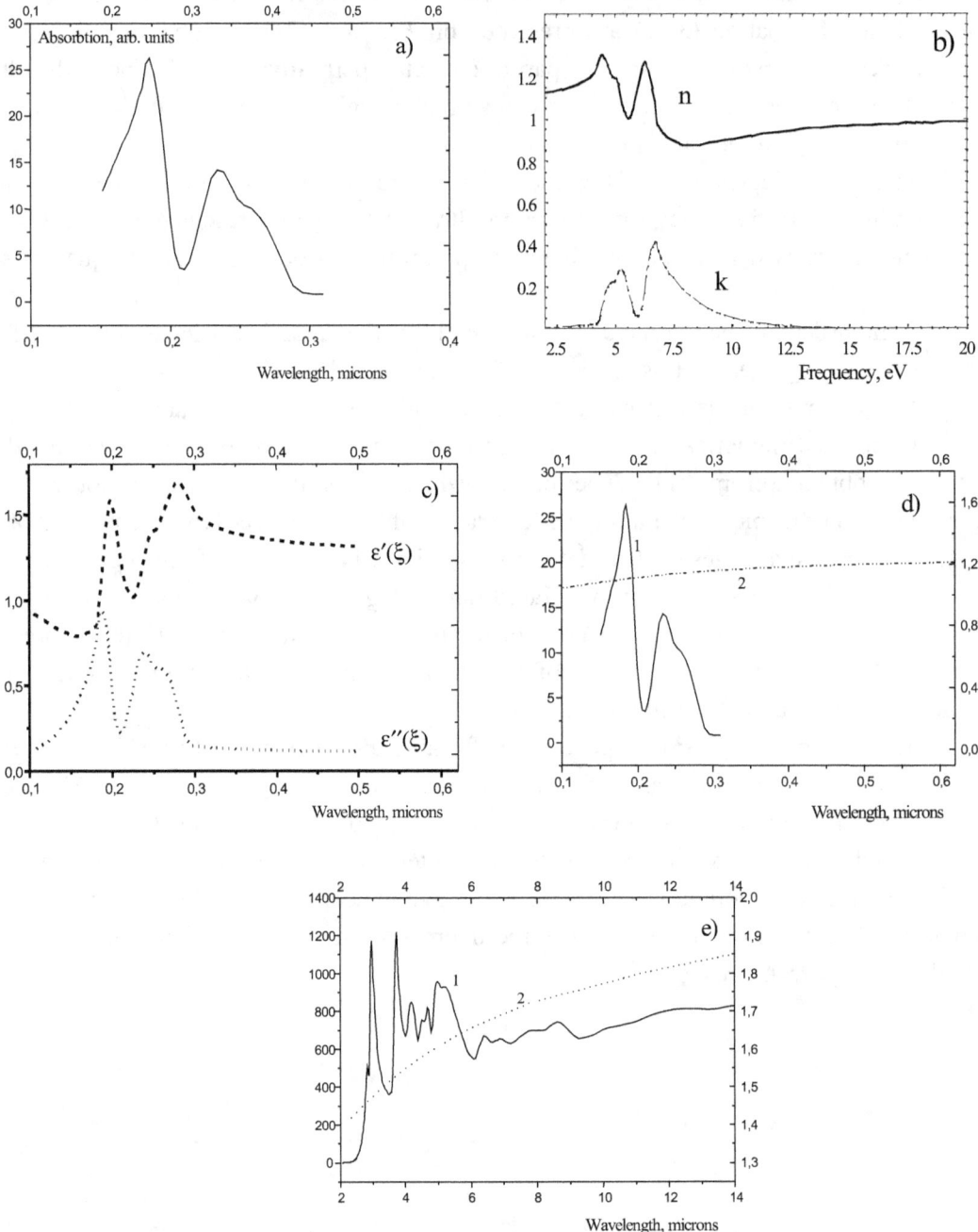

Fig. 3.3 (a) Specter of absorption of guanine in UV range (b) Dispersion dependencies of real and imaginary parts of the coefficient of refraction $n(\xi)$ and $k(\xi)$ of guanine in UV range; (c) Real $\varepsilon'(\xi)$ and imaginary $\varepsilon''(\xi)$ parts of permittivity of guanine in UV range; (d) Relation of the coefficient of absorption $\alpha(\omega)$ (curve 1) and permittivity $\varepsilon(i\xi)$ (curve 2) to the imaginary frequency for guanine in UV range; (e) Relation of the coefficient of absorption $\alpha(\omega)$ (curve 1) and permittivity to the imaginary frequency $\varepsilon(i\omega)$ (curve 2) for guanine in IR range.

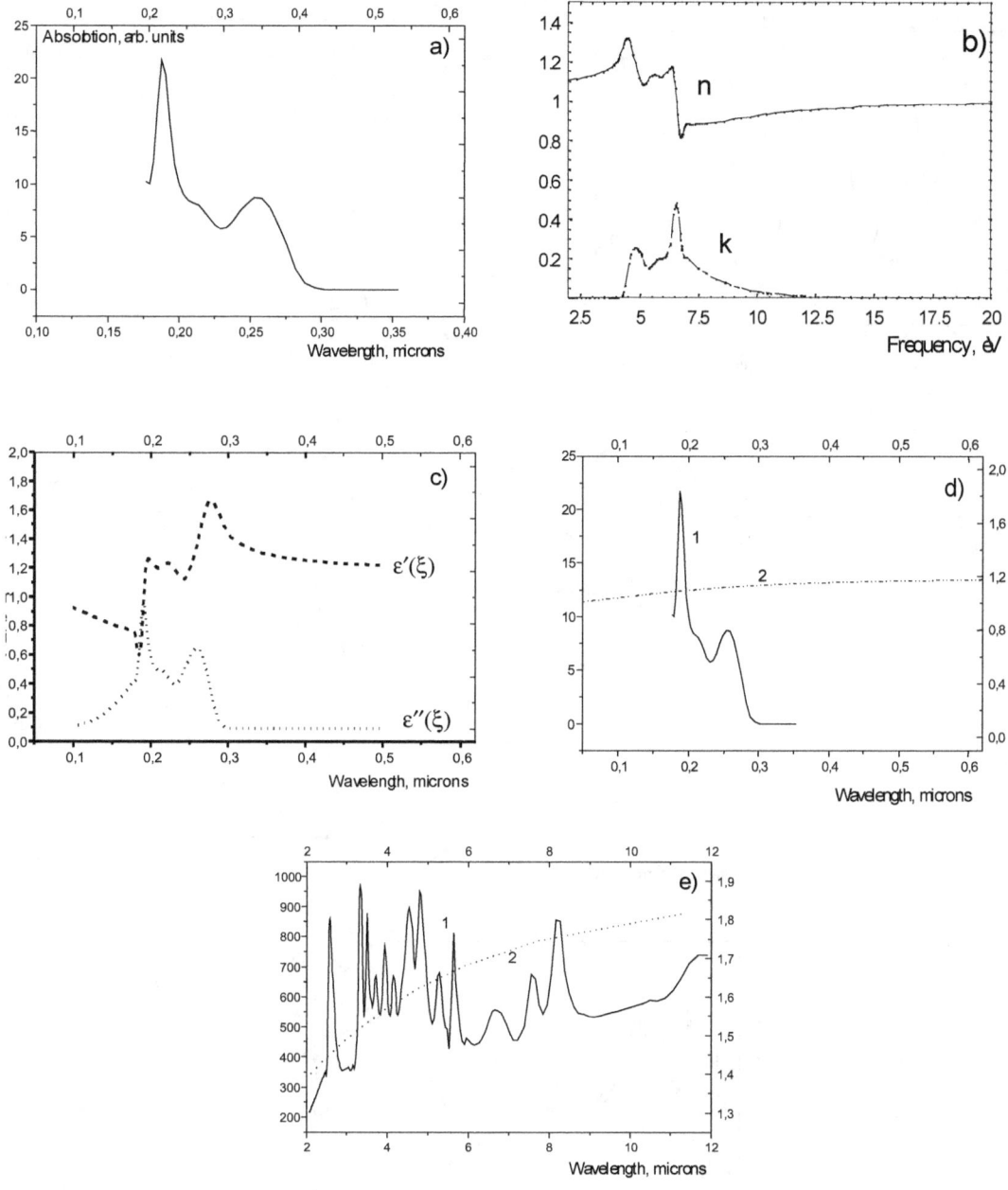

Fig. 3.4 (a) Specter of absorption of thymine in UV range (b) Dispersion dependencies of real and imaginary parts of the coefficient of refraction n(ξ) and k(ξ) of thymine in UV range; (c) Real ε′(ξ) and imaginary ε″(ξ) parts of permittivity of thymine in UV range; (d) Relation of the coefficient of absorption α(ω) (curve 1) and permittivity ε(iξ) (curve 2) to the imaginary frequency for thymine in UV range; (e) Relation of the coefficient of absorption α(ω) (curve 1) and permittivity to the imaginary frequency ε(iω) (curve 2) for thymine in IR range.

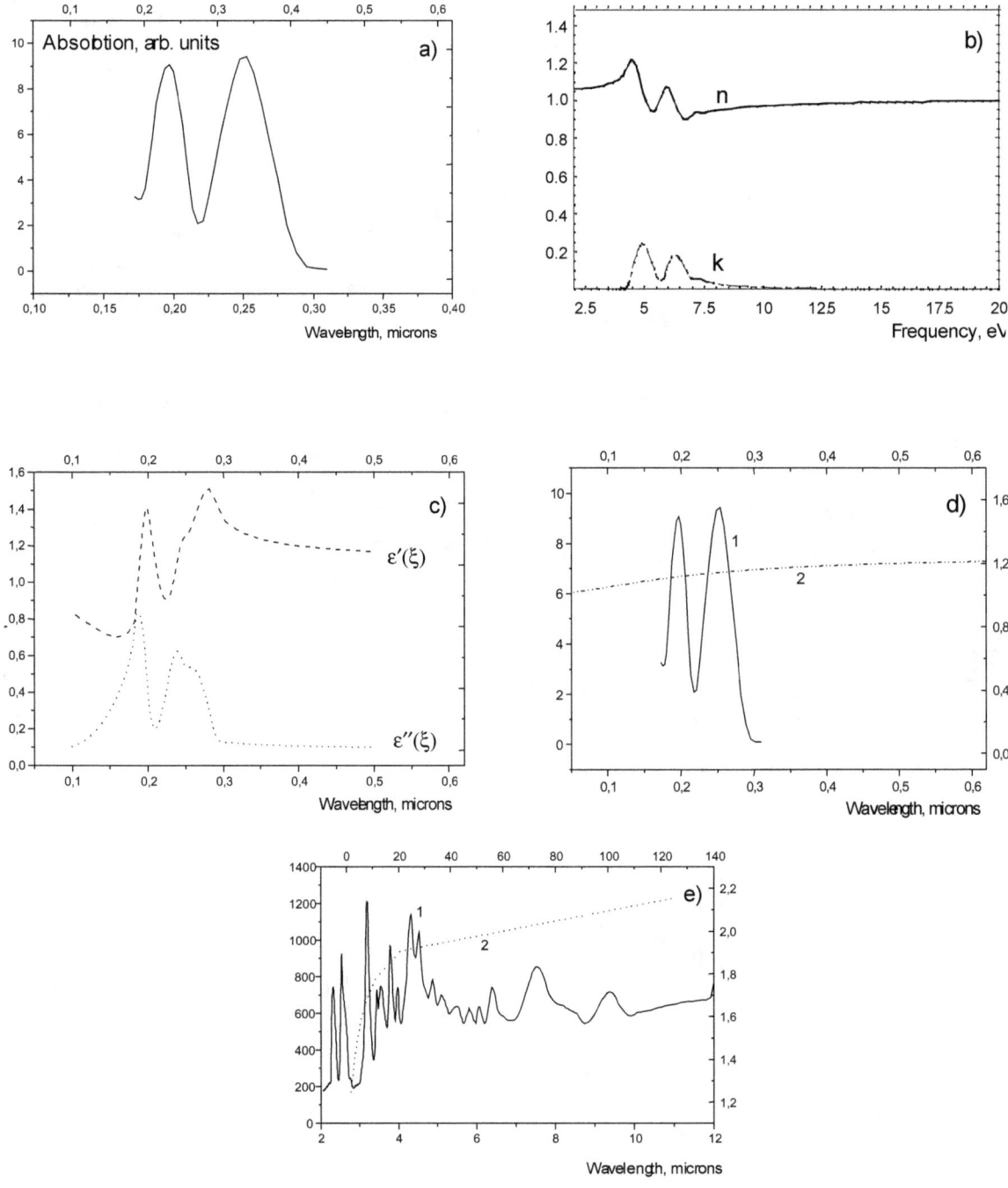

Fig. 3.5 (a) Specter of absorption of cytosine in UV range (b) Dispersion dependencies of real and imaginary parts of the coefficient of refraction n(ξ) and k(ξ) of cytosine in UV range; (c) Real ε′(ξ) and imaginary ε″(ξ) parts of permittivity of cytosine in UV range; (d) Relation of the coefficient of absorption α(ω) (curve 1) and permittivity ε(iξ) (curve 2) to the imaginary frequency for cytosine in UV range; (e) Relation of the coefficient of absorption α(ω) (curve 1) and permittivity to the imaginary frequency ε(iω) (curve 2) for cytosine in IR range.

3.3 ANALYSIS OF CHARACTERISTICS OF FORCE INTERACTION OF DNA MACROMOLECULES IN ACTIVATED WATER ENVIRONMENT WITH RADIATION AND OTHER TYPES OF DAMAGES

3.3.1 Physical features and biophysical foundations of long-distance intermolecular interaction of end pairs of nucleotides in the area of double-string DNA damages

It is a known fact that none of the enzymatic methods of reparation of damages to DNA can fully restore ruptures of complementary chains of a DNA spiral. It happens because in order for an enzyme to fulfill its "repairs" it needs a special platform, on which such restoration of DNA integrity is performed. In case of single breaks, an undamaged complimentary DNA string serves as such a platform. In case of double breaks, such possibility is no longer feasible, which, at a first glance, must lead to an irreversible destruction of DNA with a loss of genetic information. The only possibility of restoring DNA in this case is by using the method of non-enzymatic self-reparation discussed below. The efficiency of implementing the mechanism of non-enzymatic reparation depends, to a large degree, on the features of long-distance force interaction between ruptured parts of a double-string DNA spiral in liquid inner molecular environment.

In our previous works, based on modeled concepts of dispersion features of interacting nucleotides, we had discovered a non-monotonous (including repulsing as well as attracting) character of interaction between single nucleotides (Vysotskii, 1997) and their oriented pairs (Pinchuk, 2001). That interaction depended on parameters of environment and the width of a break L. In some cases, inside the area of a break (at a given critical distance L_0) there is a repulsing potential barrier $V(L_0) > 0$, affecting movement of DNA fragments. Its presence may cause rapid slowing of the process of DNA reparation and even make it completely impossible.

Let's find out, which forces work in the volume of liquid inner molecular media between damaged DNA fragments and whether there is a possibility of influencing these forces.

The full energy of interaction between end pairs of nucleotides in the area of a double break of a DNA spiral in inner molecular liquid environment consists of two additive components

$$V = V^Q + V^{vdw} \qquad (3.16)$$

The first one of them (electrostatic energy V^Q) corresponds to the sum energy of paired Coulomb interaction of charges of all atoms (Zenger, 1984) contained in the opposite nucleotides. The second component in (3.16) determines the sum dispersion electrodynamic interaction (of the kind of Van Der Waals forces) of pairs of opposing nucleotides. We shall examine both of these components separately.

3.3.2 Energy of electrostatic long-distance interaction of Watson-Creek pairs of nucleotides in the area of a double DNA break

The electrostatic component of the full energy of interaction (3.16) corresponds to the sum of paired energies of Coulomb interaction $V_{i,j}^Q$ of all charges located on the surface of the opposite end pairs of nucleotides in the area of a double break of a DNA spiral

$$V^Q = \sum_{i=1}^{N_\alpha} \sum_{j=1}^{N_\beta} V^Q_{i,j}, \qquad (3.17)$$

where N_α, N_β - number of atoms in the corresponding Watson-Creek complementary pair of nucleotides GC (guanine-cytosine) or AT (adenine-thymine);

$N_\alpha = 29$ for the GC pair, $N_\beta = 27$ for the AT pair; $V^Q_{i,j}$ – energy of Coulomb interaction of charges Q_i and Q_j located on the surface of a given Watson-Creek pair of nucleotides GC or AT.

Actual geometry of fractional charges of ions on the surface of nucleotides is presented on *Fig. 3.6* (Zaenger, 1984).

Density distribution of charges on the surface of nucleotides (*Fig. 3.7*) was calculated using the DelRe and Huckel methods, accounting for contribution of σ- and π-electrons to the total energy of interatomic link on the surface of nucleotides (Zaenger, 1984). In vacuum, electrostatic interaction of charges may be examined based on the direct use of the Coulomb rule, on the basis of actual distribution of electric charges on the surface of all nucleotides. This case precisely matches interaction in the volume of full (undamaged) DNA macromolecule. This is due to the fact, that there is no water in the space between neighboring pairs of nucleotides, which is explained by the fact that the distance of 3.4 *A* between neighboring pairs is less than diameter of a water molecule.

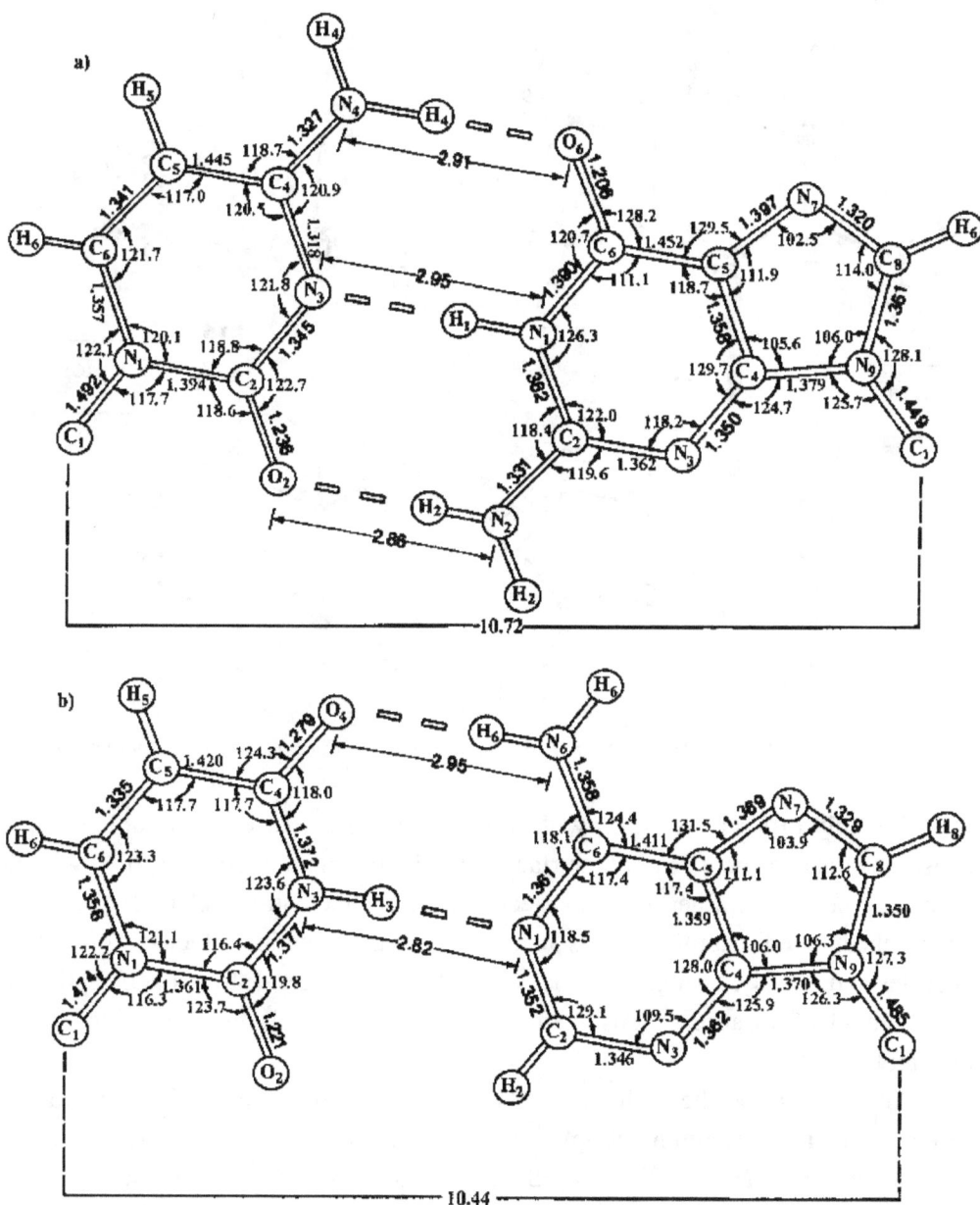

Fig. 3.6 *Coordinates of atoms for a complementary pair of nucleotides guanine-cytosine*
(a) and adenine-thymine (b). All dimensions are given in Angstrem units and angles – in
degrees.

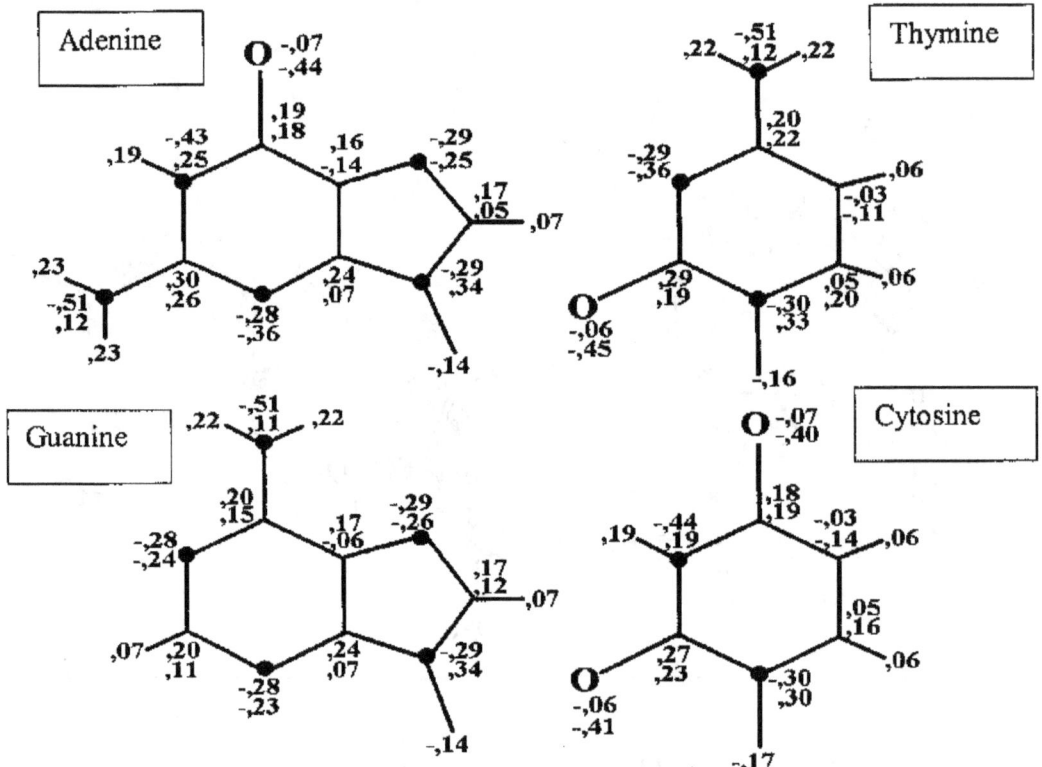

Fig. 3.7 Distribution of fractional electric charges on the surface of nucleotides

A totally different situation will present itself in the case of a double break of DNA. In that case, the distance between both surfaces of a double break would be filled with liquid water-salt media. Under the influence of an electric field of nucleotides' charges there is polarization of liquid media, redistribution of its charges, partial deflection of charges on the surface of nucleotides and, finally, changing of the result electrostatic field. It is a self-regulating process – electric field affects location of ions in a liquid media, while their distribution determines alteration of the field.

Accounting for the influence of ionic content of inner molecular water-salt media, representing an ionic solution of light mobile hydrated electrons e^- and heavy ions Na^+, K^+, Cl^-, Ca^{++}, Mg^{++}, Mn^+, H^+ and OH^- and others on efficiency of Coulomb component of interaction may be accomplished based on the Debye-Huckel theory. According to that theory finding the energy of Coulomb interaction $V^Q_{i,j}$ between fractional charges distributed on the surface of nucleotides, located in an inner cellular liquid media, is conditional on solution of the linearized equation of Poisson-Boltzmann (Landau, 1964)

$$\Delta \Psi = g^2 \Psi, \qquad (3.18)$$

for self-regulated normalized potential of electrostatic field $\Psi = e\Psi/k_b T$, altered in accordance with reverse influence of ions in a liquid media.

Here,

$$g^2 = [4\pi e^2 / \varepsilon_W(0) k_b T] \sum_z n_z z^2,$$

Ψ - self-regulated potential of electrostatic field, formed around each charge in a water-salt media, accounting for spatial redistribution of the rest of ions,

n_z – concentration of ions with the repetition factor of charge Z in the volume of inner cellular liquid (for anions $Z < 0$, for cations $Z > 0$),

$\varepsilon_W(0)$ – static value of dielectric permittivity of inner cellular water-salt media, corresponding to permanent electric field (at frequency $-\omega = 0$).

Variable g is the Debye constant, determining the radius of screening $R_s = 1/g$ of electrostatic field for any excess charge in a neutral plasma-like environment containing ions of both polarities.

Variable $n = \frac{1}{2} \sum_z n_z z^2$ is ionic power of a solution, different from zero in both charged and neutral solutions. For a neutral solution, total charge of the system is equal to zero and therefore $\sum_z n_z z = 0$.

Analytical solution of the linearized equation of Poisson-Boltzmann (3.18) may be obtained for different specific cases, including the case of interaction of two charged spheres with radiuses a_1, a_2 and effective homogeneous surface potentials $\tilde{\Psi}_1^s$, $\tilde{\Psi}_2^s$. This model precisely matches ionization of ions located on the surface of one pair of nucleotides with ions on the surface of another pair. Each of the ions can be considered a charged sphere. Fixed scalar potential in the space between these spheres in linear approximation would be expressed by the sum of potentials of separate spheres $\tilde{\Psi}_1 + \tilde{\Psi}_2$, each one of which having the form

$$\tilde{\Psi}_1(r_1) = \frac{a_1 \Psi_1^s}{r_1} e^{-g(r_1 - a_1)}$$

$$\tilde{\Psi}_2(r_2) = \frac{a_2 \Psi_2^s}{r_2} e^{-g(r_2 - a_2)} \qquad (3.19)$$

The result energy of electrostatic interaction between two charged atoms (i.e. ions on the surface of different nucleotides) located in an ionic water-salt solution has the following form (Parsegian, 1973; Pinchuk, 2001)

$$V_{i,j}^Q(R) = -\int_\infty^R \frac{\varepsilon_W(0) \, \Psi_i^s \Psi_j^s (1 + gr) a_i a_j e^{-g(r - a_1 - a_2)}}{r^2} \, dr \qquad (3.20)$$

where a_i and a_j – radiuses of interacting ions contained in the opposite nucleotides, R – distance between these ions, $\Psi_{i,j}^s$ – potentials on the surface of interacting ions.

3.3.3 Features of dispersion electrodynamic interaction of nucleotides in inner cellular media in the area of a double DNA break

The second component of the initial relation (3.16) defines total (summed) dispersion electrodynamic interaction of the type of Van Der Waals interaction

$$V^{vdw} = \sum_{i=1}^{2} \sum_{j=1}^{2} V_{i,j}^{vdw}$$

between pairs of nucleotides located on the opposite sides of a rupture of a DNA spiral. Here, $V_{i,j}^{vdw}$ – energy of dispersion interaction between separate nucleotides i and j located on the opposite sides of a double damage of a DNA spiral.

As was noted earlier in section 3.1, the energy of dispersion interaction of two objects located in a media with dielectric permittivity $\varepsilon_W(\omega)$ is defined most concisely by the general theory of Lifschitz (Dzyaloshinskii, 1961).

There are several known special cases of using this theory for actual geometry of interacting objects and the distance between them.

In particular, the energy of dispersion long-distance interaction between two particles with volumes V_i and V_j located in a media at a distance L from each other may be presented in the form of an expression (Vysotskii, 1999)

$$V_{ij}^{vdw}(L) = -\frac{27\eta}{16\pi^3} \frac{V_i V_j}{L^6} \int_0^{\infty} \frac{(\varepsilon_i(i\omega) - \varepsilon_W(i\omega))(\varepsilon_j(i\omega) - \varepsilon_W(i\omega))}{(\varepsilon_i(i\omega) + 2\varepsilon_W(i\omega))(\varepsilon_j(i\omega) + 2\varepsilon_W(i\omega))} d\omega \quad (3.21)$$

where $\varepsilon_i(i\omega)$, $\varepsilon_j(i\omega)$ – dielectric permittivity of each of the interacting particles as a function of imaginary frequency $i\omega$; $\varepsilon_W(i\omega)$ – dielectric permittivity of inner cellular water-salt media, in which particles are placed.

Expression (3.21) is true for distances $L \ll \lambda_0$, small in comparison with the wave length $\lambda_0 \approx 0.1$–0.05 micrones of the strongest atomic and molecular resonance vibrations in the volume of particles being investigated, which allows to ignore the delaying effects.

When condition $\overline{R}_{i,j} < L \ll \lambda_0$ holds, expression (3.21) directly determines the energy of dispersion interaction of nucleotides located on the ends of a double break with large distance L in an actual inner cellular liquid media with dielectric permittivity $\varepsilon_W(\omega)$.

Here, $\overline{R}_{i,j} \equiv (S_{i,j})^{1/2} \approx 7\,A$ – average linear size of each of the oriented nucleotides, presenting a group of atoms distributed along a flat surface of nucleotides and perpendicular to the direction of a DNA spiral break;

S_i, $S_j \approx 50\,A^2$ – surface areas of nucleotides located on edges of a double DNA break.

Note, that in this case total energy of dispersion interaction V^{vdw} includes the energy of interaction of each of the nucleotides belonging to a particular pair with both nucleotides on the opposite side of the break. For further analysis, it is crucial to remember that in consideration of double DNA breaks with width of breaks not exceeding $L \approx 15–20\,A$, calculation of total energy of interaction of ends of a ruptured DNA spiral may be conducted solely on the basis of investigating interactions of end pairs of nucleotides located closest to each other (i.e. without accounting for other pairs of nucleotides, further away from the area of a break). This condition owes to a very rapid decrease of contributions into the total energy of dispersion interaction from remote (at $n > 0$) pairs of nucleotides, which are inversely proportional to the distance

$$L_n = (L + n\Lambda) = L(1 + n\Lambda/L)$$

to these pairs, taken to the sixth degree (see (3.21)).

Here, $\Lambda = 3.4\,A$ – period of distribution of nucleotide pairs along the axis of the double DNA spiral.

In another extreme case of a very small distance between nucleotides $L < \overline{R}_{i,j}$, $L<<\lambda_0$, their dispersion interaction in an actual inner cellular liquid media matches that in another model solved on the basis of Lifschitz's theory'– interaction of two parallel plains separated by a thin dielectric layer with thickness L (Dzyaloshinskii, 1961).

In that model, dispersion interaction is characterized by the pressure of fluctuating electromagnetic field on the surface of these plains

$$P_{i,j} = \frac{\eta}{8\pi^2 L^3} \int_0^\infty \frac{\left(\varepsilon_i(i\omega) - \varepsilon_W(i\omega)\right)\left(\varepsilon_j(i\omega) - \varepsilon_W(i\omega)\right)}{\left(\varepsilon_i(i\omega) + \varepsilon_W(i\omega)\right)\left(\varepsilon_j(i\omega) + \varepsilon_W(i\omega)\right)} d\omega \tag{3.22}$$

For plains of limited size (S_j and S_k) the corresponding energy of interaction is characterized by the expression (Vysotskii, 1999)

$$V_{i,j}^{vdw} = -\frac{\eta}{16\pi^2} \frac{\tilde{S}_N}{L^2} \int_0^\infty \frac{\left(\varepsilon_i(i\omega) - \varepsilon_W(i\omega)\right)\left(\varepsilon_j(i\omega) - \varepsilon_W(i\omega)\right)}{\left(\varepsilon_i(i\omega) + \varepsilon_W(i\omega)\right)\left(\varepsilon_j(i\omega) + \varepsilon_W(i\omega)\right)} d\omega \tag{3.23}$$

where $\tilde{S}_N$ – actual surface area of the smaller of interacting nucleotides i or j, opposing each other in the area of a double break. For such plain-like geometry, total energy of dispersion interaction $V^{vdw} = \sum_{i=1}^{2} V_{i,i}^{vdw}$ accounts for connection of each of the nucleotides belonging to a particular pair with only a directly opposite one, on the other side of a double break.

Direct identification of the effective surface area $\mathcal{S}_N$ (in calculation of interaction between closely located plains of nucleotides) with the actual area of the smaller of nucleotides (S_i or S_j) may lead to significant errors affecting the full balance of the Coulomb energy of interaction and dispersion energy of interaction (3.16). In order to eliminate that source of errors in calculation of characteristics of dispersion interaction between oriented pairs of nucleotides in an inner cellular liquid media at the presence of a double break, a model of interaction (3.23) between pairs of nucleotides in an undamaged DNA spiral has been created and tested. During that process, using dispersion characteristics of nucleotides presented below, a value for an effective area of nucleotides S_N had been chosen, for which the estimated value of total energy of longitudinal interaction of nucleotides in an undamaged DNA spiral, based on (3.23) at the distance $L = 3.4\ A$ in the absence of water-salt media (i.e. at $\varepsilon_w(i\omega) = 1$) corresponds to the experimental value (for example, the value of 14.59 *kcal/mol*) for the dimer (guanine-cytosine)-(guanine-cytosine).

The value $\mathcal{S}_N$ was later used in calculation of dispersion interaction (3.23) for different pairs of nucleotides and their various mutual orientations at the presence of a double break and inner cellular liquid media.

As further analysis shows, the range of distances between nucleotides (the width of a double break) approaching the critical value $L \approx \overline{R}_{i,j}$, which separates areas, where expressions (3.21) and (3.23) could be used, is the most interesting. For such a range of values for L, none of the solutions for (3.21) and (3.23) is adequately correct.

To find the explicit form of the energy of dispersion interaction on the entire range of distances L (including the area $L \approx \overline{R}_{i,j}$) straight interpolation, using the method of splines of the third order, was used. Then, calculated expressions (3.21) and (3.23) for intervals $L \geq 2\ \overline{R}_{i,j}$ _ $L \leq 2\ \overline{R}_{i,j}$, where these expressions could be applied, were "bound" in an intermediate interval $\overline{R}_{i,j}/2 \leq L \leq 2\overline{R}_{i,j}$, where both formulae (3.21) and (3.23) are not true.

Fig. 3.8 shows relation of the energy of dispersion interaction of opposing dimers (guanine-cytosine)-(guanine-cytosine) calculated on the basis of the expression (3.21), true for the area of relatively large distances, as well as the expression (3.23), which is true in the interval limited by very small distances. The diagrams show, that calculations made on the basis of (3.21) produce a very big error for the energy of interaction in the area of small distances, while calculations made on the basis of (3.23) give a big error for large distances. The same figure has the interpolation curve between these relations into the area of medium distances $\overline{R}_{i,j}/2 \leq L \leq 2\overline{R}_{i,j}$, obtained by using the method of splines (Pinchuk, 2001).

We can see from this figure, that interpolation of obtained relations allows us to determine the explicit form of the expression for the energy of dispersion interaction in the area, where an adequate theoretical model does not exist. The result of such interpolation is used later for calculation of total energy taking Coulomb electrostatic interaction into account.

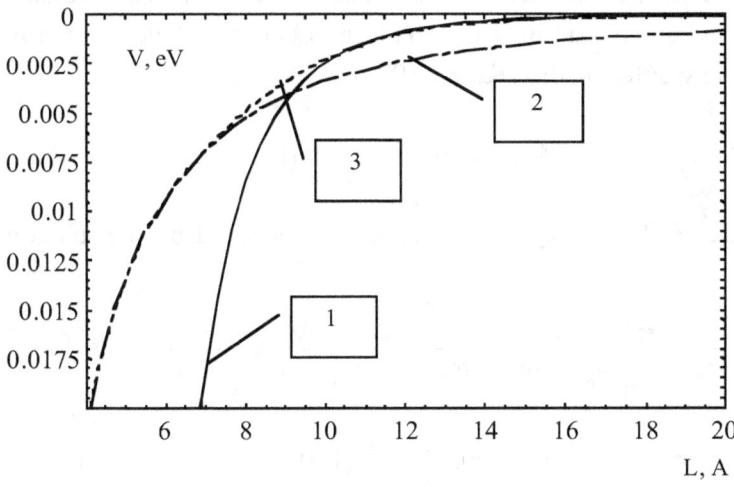

Fig. 3.8 *Energy of Van Der Waals interaction between pairs (guanine-cytosine)- (guanine-cytosine) in relation with distance L between them. The solid curve 1 is created on the basis of relation (3.21), the line-dotted curve 2 is represents the relation (3.23), the chain curve 3 represents interpolation by third-order splines.*

This approach makes it possible to calculate the energy of Van Der Waals interaction of opposing pairs of nucleotides with high precision in the interval of possible values for width L of a double DNA break, which is most interesting from the point of view of stability of molecules of DNA.

3.4   NUMERIC MODELING OF THE ENERGY OF INTERACTION OF END PAIRS OF NUCLEOTIDES IN THE AREA OF A DOUBLE-STRING DNA RUPTURE

Based on the relations obtained earlier for Coulomb and dispersion energies of interaction of end pairs of nucleotides in the area of a double-string DNA rupture, all components of the sum energy of interaction were calculated. In order to calculate the integrals in (3.21) and (3.23), determining the energy of Van Der Waals interaction between opposing nucleotides, calculation results for dispersion of dielectric permittivity for all four types of nucleotides in ultraviolet and infrared ranges of the specter mentioned earlier, were used. Then, using the method of splines, extrapolation of obtained values for dielectric permittivity was performed into the range of frequencies between the infrared and ultraviolet frequency ranges, where intensive line of nucleotide absorption is absent.

Such extrapolation was also made into the area of frequencies, lying between the ultraviolet and Roentgen ranges. In the Roentgen range, the structure of the expression for dielectric permittivity of nucleotides matches the standard "plasmatic" approximation and is determined by the expression (3.10).

Dispersion of dielectric permittivity $\varepsilon_w(\omega)$ of inner cellular water-salt medium was modeled with the help of the expression, determining the connection between dielectric permittivity $\varepsilon_w(\omega)$ and sensitivity $\chi(\omega)$ for any solid media (Kittel, 1970)

$$[\varepsilon_w(\omega)-1]/[\varepsilon_w(\omega)+2] = 4\pi\chi(\omega)/3[1+4\pi\chi(\omega)/3] \qquad (3.24)$$

In (3.24), sensitivity of water $\chi(\omega) = n_w\alpha(\omega)$ is determined by the relation

$$4\pi\chi(\omega) = \frac{n_w\alpha_0(\omega)}{1-i\omega\tau_r} + \sum_{K=1}^{4}\frac{\Delta\varepsilon_K\omega_K^2}{\omega_K^2 - \omega^2 - i\omega\Gamma_K} + \frac{\Delta\varepsilon_{UV}\omega_{UV}^2}{(\omega_{UV}^2 - \omega^2 - i\omega\,\Gamma_{UV})} - \omega_p^2/\omega^2 \qquad (3.25)$$

This relation follows from expressions (1.1) – (1.5) for polarizing ability of water $\alpha(\omega)$ and its relation to frequencies in different parts of the electromagnetic specter, discussed in chapter 1.

Calculation of Coulomb electrostatic interaction required the use of published values for charges distributed on surfaces of corresponding nucleotides (Zenger, 1997). These values are presented on *Fig. 3.7*. Distribution of these charges is the result of a complicated process of redistribution of electronic density during formation of covalent ionic links, included in certain nucleotides. Coordinates of atoms in nucleotides carrying fractional charges are presented on *Fig. 3.6*.

On *Fig. 3.9* there is relation of electrostatic energy of interaction of oriented pairs of nucleotides *GC-GC* from distance between them, calculated with the help of the expression (3.20). The same figure shows the relation, calculated with the help of expressions (3.21) and (3.23), of dispersion electrodynamic energy of interaction of the same pairs of nucleotides from the distance between pairs L, as well as the expression (3.16) for sum energy of interaction.

We can see from the provided results that the electrostatic energy of Coulomb interaction corresponds to a strong mutual repulsion of pairs of nucleotides. This energy causes a sharp increase of repulsion, as the distance becomes smaller. Apparently, the character of relation of this energy from distance is monotonous, with constant polarity.

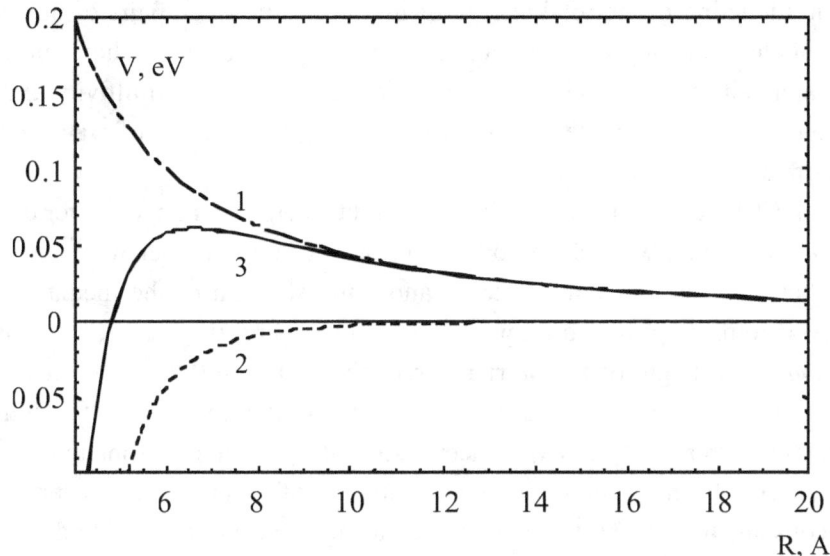

Fig. 3.9   The energy of electrostatic interaction (curve 1), energy of electrodynamic interaction (curve 2) and total energy of interaction (curve 3) between pairs (guanine-cytosine)-(guanine-cytosine) and their relation to the distance between them L.

On the other hand, the electrodynamic dispersion energy of interaction corresponds to mutual attraction, provided that with smaller distance such attraction becomes stronger. Increasing of attraction, which is equivalent to reduction of negative potential energy of interaction) also complies with the monotonous rule. Increasing of dispersion energy of attraction stops only at such a small distance (about 1 A), when electronic coats of atoms overlap on the surface of nucleotides and repulsion caused by the Pauli rule begins to have an effect.

Each one of these forces characterizes a specific type of interaction (a potential barrier or a potential pit on the whole range of distances between pairs of nucleotides on the opposite ends of a double break).

Contrary to that, the sum energy of interaction (curve 3 on *Fig. 3.9*, which is formed because of incomplete mutual compensation of forces of repulsion and attraction, is a non-monotonous function of width of a double break L between oriented pairs of nucleotides (Pinchuk, 2001), with varying polarity. From this relation it follows, that long-distance interaction between these pairs in a defined interval of values $L \approx 7$ A results in appearance of a potential barrier $V(L) > 0$, with height of about 0.07 eV, which corresponds to $V_{max} \approx 3$ $k_B T$. These results were obtained at normal (physiologically) viscosity of the inner cellular media, with $g = 0$ and relative concentration of hydrated electrons (in relation to concentration of water molecules n_w) in a water-salt media $\alpha < 10^{-9}$.

Since the height of this barrier is considerably higher than the energy of thermal movement, equal to $k_B T \approx 0.025$ eV at the room temperature; its presence may significantly affect the dynamics of movement of ruptured DNA fragments. In particular, this potential barrier may radically alter the probability of the self-reparation process of DNA following a double rupture of a DNA spiral. Such occurrences will be discussed in full in the next chapter.

Numeric modeling of the total energy of interaction between pairs of nucleotides shows, that the height of the repulsing anti-reparation barrier V_{max} depends on the ionic content of the inner cellular water-salt media, its viscosity, temperature, dielectric permittivity and several other parameters. Some of those parameters may change during the process of water activation. Let's consider some of those relations explicitly.

On *Fig. 3.10* there are calculated relations of total energy of interaction for dimers (guanine-cytosine)-(guanine-cytosine) with physiologic viscosity of inner cellular water-salt media $\eta = \eta_0$ to the distance between pairs of nucleotides L and ionic strength of the media.

It follows from the provided graph that with increasing the value of the Debye constant to $g = 2.5*10^7 \; sm^{-1}$, the height of the barrier is reduced to $V_{max} \approx 0.9 \; k_B T$, when its influence on the process of self-reparation is considerably weaker. Similar reduction of height of the anti-reparation barrier corresponds to relative concentration of hydrated electrons in a water-salt media on the level of $\alpha = 10^{-8}$. Further increasing of concentration of ions in a water-salt media (increasing of the Debye constant to $g = 5*10^7 \; sm^{-1}$ or increasing concentration of hydrated electrons to $\alpha = 10^{-7}$) causes a potential barrier to disappear, which is compatible with mutual attraction of ruptured DNA fragments at any distance.

Increasing solution's viscosity to $\eta = 10\eta_0$ results in reduction of height of the anti-reparation barrier to $V_{max} \approx 2 \; k_B T$ for a pair (guanine-cytosine)-(guanine-cytosine) (*Fig. 3.11*) with the same relation to the Debye constant and concentration of hydrated electrons, as with physiologic viscosity.

Fig. 3.12 shows the results of numeric calculations of the total energy of interaction for dimers (guanine-cytosine)-(guanine-cytosine) at a very high viscosity of the inner cellular media $\eta = 100\eta_0$. In this case, the height of the barrier is even lower, at $V_{max} \approx 0.6 \; k_B T$.

Analogous repulsing barrier with the height $V_{max} \approx 1.5 \; k_B T$ at physiologic viscosity of solution $\eta = \eta_0$ appears at the same distance L between pairs of nucleotides (adenine-thymine)-(adenine-thymine) and disappears with increasing the Debye constant to $g = 10^7 \; sm^{-1}$ (*Fig. 3.13*).

Increasing the viscosity to $\eta = 10\eta_0$ results in reduction of the height of the anti-reparation barrier for interaction between these pairs (see *Fig. 3.14*), with preservation of the same relation to the ionic content of a liquid media as with physiologic viscosity.

On *Fig. 3.15* and *3.16* there is calculation of influence of additional long-distance interaction between sugar-phosphate attached groups, connecting separate pairs of nucleotides into a single periodic system of a DNA macromolecule, and located on either the same or different sides of a double break. We can see, that accounting for such an interaction makes the parameters of the anti-reparation barrier somewhat more accurate, while leaving all the main patterns unaffected.

The barrier is not present between the rest of the possible combinations of oriented pairs of nucleotides (adenine-thymine)-(guanine-cytosine), (adenine-thymine)-(cytosine-guanine), (adenine-thymine)-(thymine-adenine), (guanine-cytosine)-(cytosine-guanine), located on the ends of a double DNA break, and total energy of interaction of pairs of nucleotides is negative ($V(L) < 0$) at any width of a double break of DNA macromolecules. This result supports the fact that total interaction given this form of a double break always corresponds to mutual attraction of fragments of DNA.

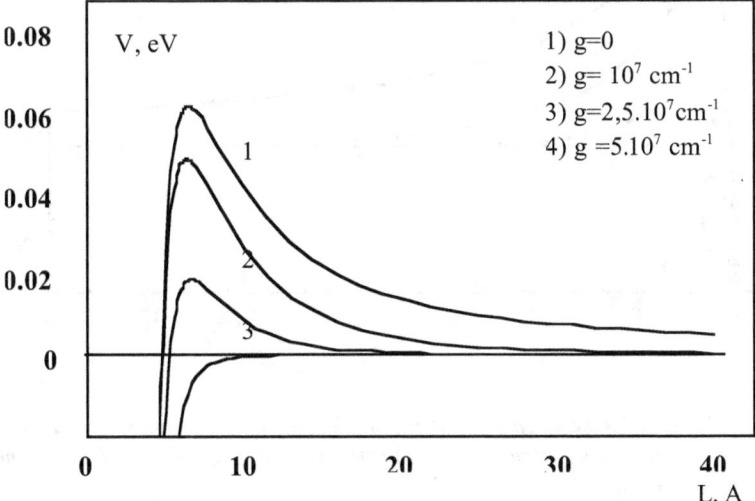

Fig. 3.10   *Total energy of interaction between pairs of nucleotides (guanine-cytosine)-(guanine-cytosine) located on the opposite ends of a ruptured DNA spiral at natural viscosity of the inner cellular physiologic solution.*

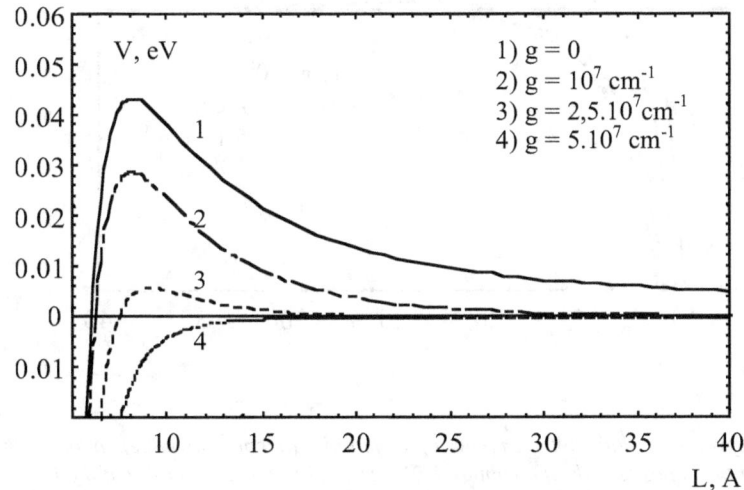

Fig. 3.11   *Total energy of interaction between pairs of nucleotides (guanine-cytosine)-(guanine-cytosine) located on the opposite ends of a ruptured DNA spiral at a higher viscosity of the solution $\eta = 10\eta_0$*

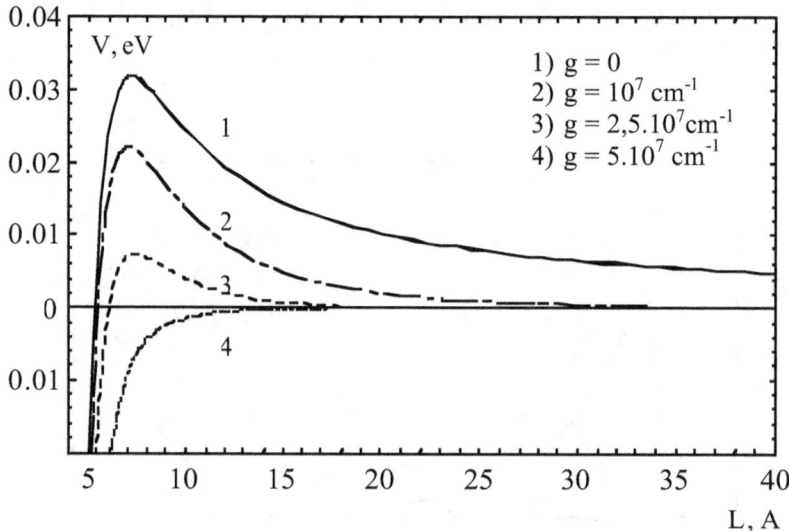

Fig. 3.12   Total energy of interaction between pairs of nucleotides (guanine-cytosine)-(guanine-cytosine) located on the opposite ends of a ruptured DNA spiral at a very high viscosity of the inner cellular media $\eta = 100\eta_0$

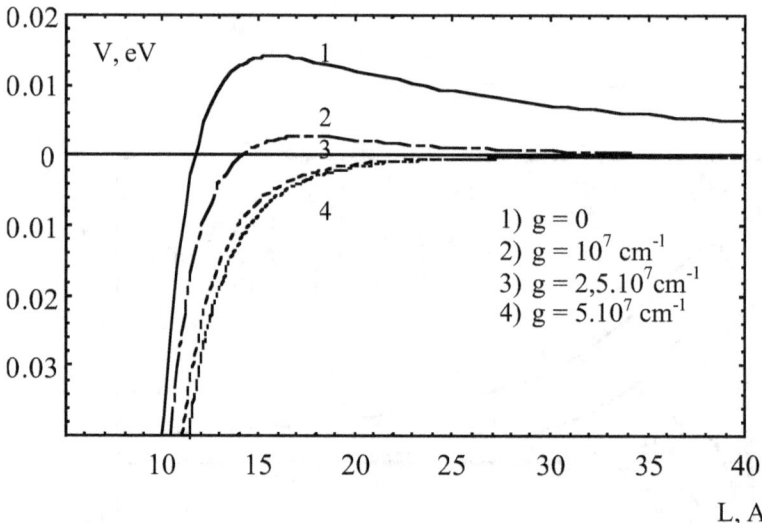

Fig. 3.13   Total energy of interaction between pairs of nucleotides (adenine-thymine)-(adenine-thymine) located on the opposite ends of a ruptured DNA spiral at natural viscosity of the inner cellular media $\eta = \eta_0$ (viscosity of a physiologic solution).

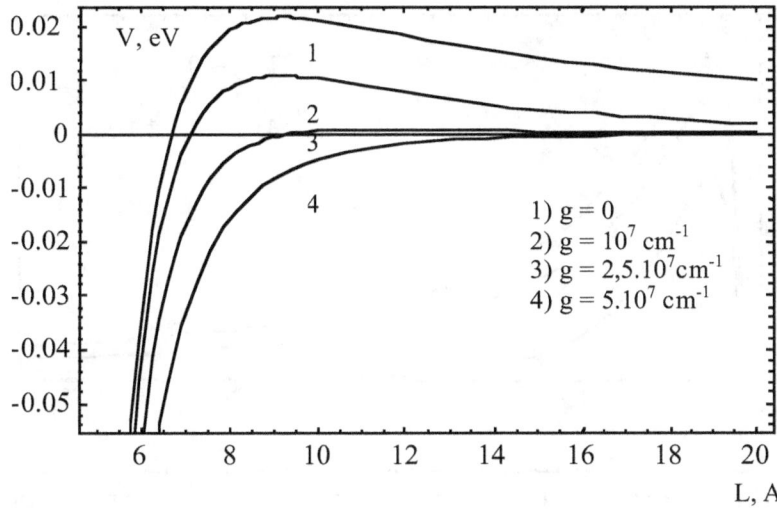

*Fig. 3.14   Total energy of interaction between pairs of nucleotides (adenine-thymine)-(adenine-thymine)
located on the opposite ends of a ruptured DNA spiral at higher viscosity of the inner cellular
water-salt media $\eta = 10\eta_0$*

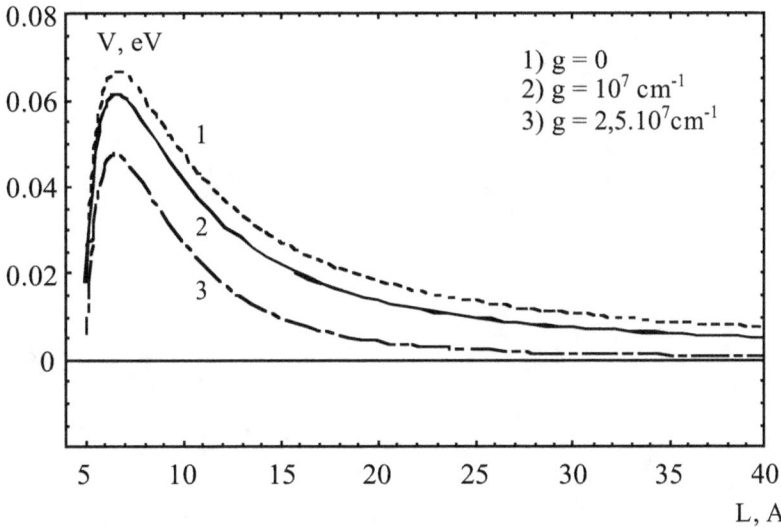

*Fig. 3.15   Total energy of interaction between pairs of nucleotides (guanine-cytosine)-(guanine-cytosine)
located on the opposite ends of a ruptured DNA spiral, accounting for additional long-distance
interaction between sugar-phosphate attached groups located on the same side of a double break.*

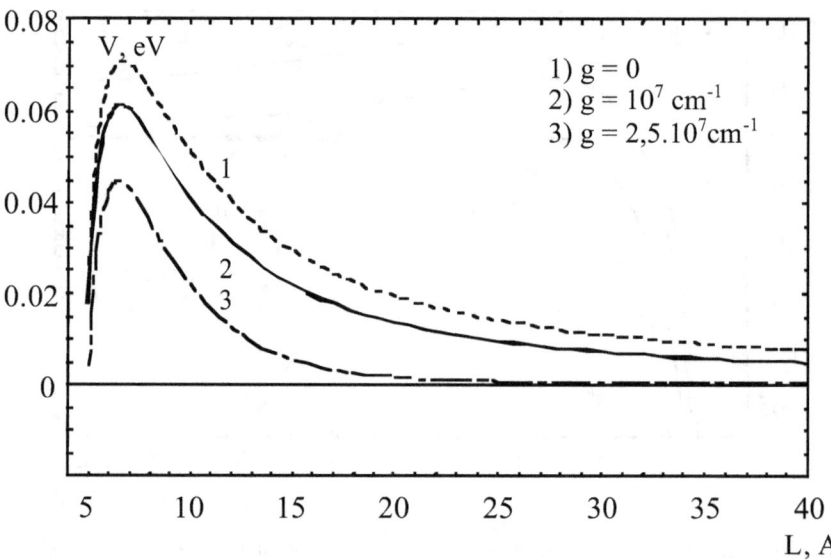

Fig. 3.16   *Total energy of interaction between pairs of nucleotides (guanine-cytosine)-(guanine-cytosine)
located on the opposite ends of a ruptured DNA spiral, accounting for additional long-distance
interaction between sugar-phosphate attached groups located on the opposite ends of a double break.*

Literature to Chapter 3

Alberts B. et. al., Molecular biology of a cell, Moscow, 1994

Barash Yu. S., Van-der-Vaals Forces, Nauka, Moscow, 1988

Dzialoshinskii I.E., Lifshitz E.M., Pitaevsky L.P.// Soviet Physics - Uspekhi, v 53 (1961) *p. 381.*

Kantor George J., "DNA Damage and Repair", Molecular Biology and Biotechnology, NY, VCH Publishers, Inc. 1995, *p. 217*

Kittel Charles., Introduction to Solid State Physics, John Willey and Sons. Inc., 1970.

Kudryashev Yu.S., Berenfeld B.S. Foundation of radiation biophysics, Moscow, Moscow State Univ. Publ. House, 1982.

Landau L.D., Lifshitz E.M. Electrodynamics of continuous mediums, Moscow, 1959.

Landau L.D., Lifshitz E.M. Statistics Physics, Moscow, 1964.
London F. // Z. Phys., v. 63 (1930) *p. 245.*

London F., Eisenshitz F. // Z. Phys., v. 60 (1930) *p. 491.*

Parsegian V.A., Physical Chemistry: Enriching Topics From Colloid and Surface Science, La Jolla, THEOREX, 1975, *72 p.*

Pinchuk A.O., Vysotskii V.I. // Physical Review E, v. 63, (2001) *p. 31904.*

Pinchuk A.O., Vysotskii V.I., Zhmudskii A.A. Spectroscopy of Biological Molecules: New Directions, Kluwer Academic Publisher, 1999, *p. 403*

Pinchuk A.O., Vysotskii V.I. // Bioelectrochemistry and Bioenergetics, Elsevier, v. 48, (1999) *p. 329.*

Ryabchenko M.I. Radiation and DNA, Atomizdat Publishing House. Moscow, 1979 (in Russian).

Sabisky E.S., Anderson C.H. // Phys. Rev. A, v. 7, (1973) *p. 790*

Vysotskii V.I., Kornilova A.A., Pinchuk A.A., Scherbakov L.V. // Defensive Technique, #6 (2003) *p.16* (In Russian).

Vysotskii V.I., Pinchuk A.A., Kornilova A.A., Samoylenko I.I. //Nuclear and Radiation Safety, v. 2, (1999) *p. 66* (In Russian).

Vysotskii V.I., Pinchuk A.A., Kornilova A.A., Samoylenko I.I. // Radiation Biology. Radio-ecology, v. 37 (1997) *p. 494.*

Young L., V.V. Prabhu and E.W. Prohofsky // Phys. Rev. A, v. 39 (1989) *p. 3173*

Zaenger W., Principles of Nucleic Acid Structure, New York: Springer-Verlag, 1984.

4. ACTIVATED WATER AND THE ISSUE OF LIFE SAVING STABILITY OF BIOLOGICAL SYSTEMS UNDER INFLUENCES OF LIFE SUPPRESSING CONDITIONS

4.1 EQUATIONS OF NON-STATIONARY DYNAMICS OF CREATION AND ELIMINATION OF THE DOUBLE-STRAIN DNA BREAKS FOR COMBINED IONIZING IRRADIATION AND THE EFFECT OF FREE RADICALS IN WATER ENVIRONMENT

4.1.1 Dynamics of double-strain DNA breaks in thermodynamically balanced water environment

In chapter 3 we have demonstrated that the pattern of interaction between end pairs of nucleotides (adenine (A), thimine (T), guanine (G), cytosine (C)) located on both sides of a double break of a DNA spiral, intricately depends on the width of a break L, dispersion of dielectric permittivity $\varepsilon_w(\omega)$ and viscosity η of the inner cellular media, as well as spatial orientation, dielectric permittivity $\varepsilon_i(\omega)$ and charge characteristics of these pairs of nucleotides. That interaction includes both the sum of Coulomb interactions of different pairs of nucleotides, distributed on the surface of charges, and Van Der Waals electrodynamic dispersion interaction between these pairs in the inner cellular media. Based on the performed analysis, it was discovered, that for any two combinations of end pairs of nucleotides (GC–GC) and (AT–AT) with the width of a break about $L_0 \approx 7\ A$, there is a potential barrier $V(L) > 0$. It resists the process of reparation of the breaks with the width exceeding L_0 (*Fig. 4.1*).

For other possible combinations of end pairs (AT–TA, AT–GC, AT–CG, GC–CG) the pattern of interaction between broken parts of DNA at any width of breaks L corresponds to their mutual attraction with a possibility of subsequent self-reparation of a double break and restoration of DNA integrity. The height of that anti-reparation barrier $V_{max} \equiv V(L_0)$ depends on the types of nucleotides in end pairs and characteristics of the inner cellular media (viscosity η, temperature T, ionic and salt content, concentration of quasi free electrons α and other factors). The value V_{max} may exceed thermal energy k_bT, but it is significantly lower than the general depth of a potential pit, which characterizes stable connection of pairs of nucleotides in a DNA chain (the *stacking* energy). For example, $V_{max} \approx 3\ k_bT$ for end pairs of nucleotides GC–GC. Meanwhile, the depth of a potential pit for a stable longitudinal interaction of the same pairs of nucleotides GC–GC has the value $V_s \approx 15\ k_bT$, which satisfies the condition $V_{max}/V_s \ll 1$.

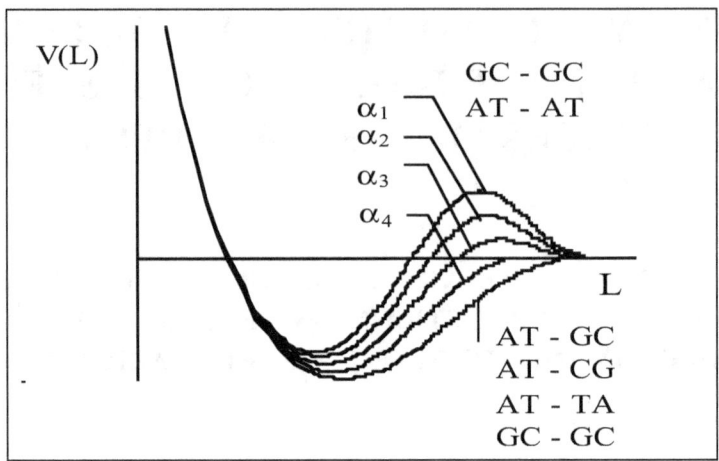

Fig. 4.1   Dependence of the spatial structure of the anti-reparation barrier in a long-distance interaction of end pairs of nucleotides in the area of a DNA break on concentration of hydrated electrons and disturbed atoms: $\alpha_1 < 10^{-9}$; $\alpha_2 = 10^{-9}$; $\alpha_3 = 10^{-8}$; $\alpha_4 \geq 10^{-7}$

Given the effect of irradiation during radiolysis, considered in the Introduction, there is a change of the ionic content and other characteristicsof the inner cellular liquid (including the Debye constant and concentration of hydrated electrons).

It was demonstrated in 3.4 that these changes lead to violation of balance between Coulomb and Van Der Waals electrodynamic dispersion interaction between nucleotides on both sides of a double DNA break, which, in its turn, causes deformation of the anti-reparation barrier $V(L)$ and reduction of its height V_{max}. Specifically, the influence of hydrated electrons in water environment on the height of a potential barrier V_{max} begins at $\alpha \approx 10^{-9}$, while at $\alpha \approx 10^{-7}$ the barrier disappears (with $V(L) \leq 0$). Elimination of potential barrier leads to a higher probability of self-reparation of double breaks and suppression of processes of DNA degradation.

Let's note another circumstance. Since the main contribution to Van Der Waals interaction is determined by the spectral area of frequencies belonging to ultraviolet range, a comparable contribution to the given effect of stimulated reduction of the anti-reparation barrier can be made by both the hydrated electrons and other quasi free electrons – for example, those in disturbed atoms.

For these electrons, the energy of connection corresponds to areas of visible, infrared and microwave frequency ranges, i.e. frequencies that are significantly lower than the frequency of the ultraviolet range. For that reason, during interaction of the fluctuating ultraviolet radiation with such electrons, they may be considered quasi free.

This is a normal approach for quantum electrodynamics. For instance, during interaction of hard Roentgen radiation with atomic electrons it is customary to treat all these electrons as free.

Based on this assumption, by the term "hydrated electrons" we shall hence mean all free and quasi-free electrons in the inner cellular media.

Dependence of the height of a potential barrier $V_{max} \equiv V(L)$ from relative concentration of hydrated electrons a corresponds to a linear approximation with a good degree of precision

$$V_{max} \approx V_0(1 - G\alpha n_W), \ \alpha \leq 1/Gn_3; \ V_{max} \approx 0, \ \alpha > 1/Gn_W \qquad (4.1)$$

Here: n_W – concentration of water molecules in inner cellular water-salt media, $G \approx 10^7/n_W$ – approximating parameter of calculations (Vysotskii, 2002).

Let's consider changing of α in time under the influence of external ionizing fields, accounting for the process of permutation of the hydrated electrons with heavy ions. In order to account for the result of simultaneous effect of several types of ionizing radiation on a biological object in the future, we shall consider the influence of two different ionizing fields on the process of formation of these hydrated electrons.

Relative concentration α of quasi free electrons in the volume of inner cellular liquid in the presence of two types of radiation can be found from the balanced relationship (Vysotskii, 2002)

$$d\alpha/dt = [(\beta_{1e} + \beta_1^*)\sigma_{13}J_1 + (\beta_{2e} + \beta_2^*)\sigma_{23}J_2] + w_e - \alpha/\tau_e \qquad (4.2)$$

where w_e – probability (in a unit of time) of non-radiation ionization of one molecule of water; τ_e – average life time (until recombination) of a hydrated electron and an excited atom.

We can see from the experiments that the average life time of a hydrated electron in clean water is equal to $\tau_e \approx 0.72 \ ms$. Presence of a salt component in water slightly changes that time. Average life time of an excited atom in ultraviolet and harder frequency ranges in water equals $\tau_e \approx 10–100 \ ns$.

Items in the right part of equation (4.2) have the following meaning. The first two items determine an increase of concentration of quasi free electrons due to disturbance of atoms and their ionization, the third item describes the process of generation of the same electrons due to non-radiation mechanism of ionization of molecules of inner cellular liquid (for example, due to thermal ionization or the effect of mechanisms of generation of free radicals independent of radiation), the last item characterizes processes of relaxation and recombination of atoms.

The mechanism of non-radiation generation of hydrated electrons is a permanent factor determining, in essence, the initial balanced concentration of these electrons

$$\alpha(0) = w_e\tau_e \qquad (4.3)$$

The initial balanced concentration can be determined from (4.2) provided that the system $d\alpha/dt = 0$ is stationary in case of absence of external ionizing fields (at $J_1 = J_2 = 0$).

Using expression (4.3), we find the final form of the equation for estimating relative concentration of hydrated electrons α

$$d\alpha/dt = [(\beta_{1e} + \beta_1^*)\sigma_{13}J_1 + (\beta_{2e} + \beta_2^*)\sigma_{23}J_2] + [\alpha(0) - \alpha]/\tau_e \qquad (4.4)$$

Here: σ_{13}, σ_{23} – cross sections of photo absorption by water molecules of two different types of radiation with intensities J_1 and J_2 respectively; $1/\beta_{(1,2)e}$ and $1/\beta_{(1,2)}^*$ – average energy spent on generation of a hydrated electron or excited atom under the influence of these types of radiation.

From (4.4), we find the expression for concentration of quasi free electrons under the influence of non-stationary ionizing fields $J_1(t)$ and $J_2(t)$ on biological system

$$\alpha(t) = \alpha(0) + \int_0^t [(\beta_{1e}+\beta_1^*)\sigma_{13}J_1(t_1) +(\beta_{2e}+\beta_2^*)\sigma_{23}J_2(t_1)]exp[-(t-t_1)/\tau_e]dt_1 \quad (4.5)$$

From the structure of the expression for $\alpha(t)$, it is evident that the value of $\alpha(t)$ strongly depends from relation between the value τ_e and characteristic time Δt of changing intensity of ionizing fields J_1 and J_2.

This expression becomes much simpler if a biological system with DNA is irradiated by a slowly changing (over the time interval Δt_1, which is much higher than average time of relaxation τ_e) ionizing fields

$$\alpha(t) \approx \alpha(0) + [(\beta_{1e} + \beta_1^*)\sigma_{13}J_1(t) + (\beta_{2e} + \beta_2^*)\sigma_{23}J_2(t)]\tau_e \quad (4.6)$$

Balanced concentration of quasi free electrons in case of constant intensity of ionizing fields $J_{1,2}(t) = J_{1,2} = $ const has the same form

$$\alpha(t) \approx \alpha(0) + [(\beta_{1e} + \beta_1^*)\sigma_{13}J_1 + (\beta_{2e} + \beta_2^*)\sigma_{23}J_2]\tau_e \quad (4.7)$$

If duration of irradiation Δt_1 is lower than the average life time of a hydrated electron τ_e, then

$$\alpha(\tau) \approx \alpha(0) + exp(-t/\tau_e) [(\beta_{1e} + \beta_1^*)\sigma_{13}J_1(t) + (\beta_{2e} + \beta_2^*)\sigma_{23}J_2(t)]\tau_e \quad (4.8)$$

In this case $\alpha(t) \rightarrow \alpha(0)$ at $t \gg \tau_e$.

Let's consider the dynamics of formation, stabilization and possible elimination of double breaks of DNA under the influence of both external combined radiation by several radiation fields and free radicals of non-radiation origin with combined duration exceeding τ_e.

The process of initial formation of a double break in a double-string DNA considering all temporal and spatial characteristics is a very complicated dynamic problem, in which both the movement of water molecules and static patterns of these molecules' effect on DNA should be taken into account. We begin at the point, where, due to an external influence on a DNA spiral, a double break with a certain width L appears. In order to make the analysis of special features of a combined effect of several types of radiation (at all possible variations of parameters of different types of radiation) more visual and explicit, we shall limit ourselves by considering the case, when the width of a break exceeds the critical point $L_0 \approx 7\ A$, signifying the existence of a natural anti-reparation barrier with height V_{max}. With a lower initial width of a break, it is usually eliminated due to combined effect of electrostatic and dispersion forces.

The pattern of such process should account for a maximum number of different combinations of external influence on DNA: effect of a separate ionizing radiation J; effect of a separate non-radiation factor of degradation W (for example, influence of free radicals); simultaneous effect of any combination of non-radiation impact and different types of radiation (J_1 and J_2) with varying efficiencies of formation of double breaks (see *Fig.4.2*). Since any break with a determined width is thermodynamically unstable (it may increase or decrease), the pattern should also consider processes of relaxation of DNA breaks.

Such processes can in two mutually opposing directions.

One part of relaxation processes leads to reversible relaxation of double DNA breaks. Reversible relaxation characterizes such a movement of DNA fragments, when a break is autonomously reduced (up to its total elimination) due to long-distance interaction of these fragments and absence of anti-reparation barrier with height V_{max} is characterized by phenomenological probability $1/T_1$ representing transition from condition of reversible double break to the state of whole and intact DNA. This phenomenon is caused by reduction of width of a double break L to a size comparable with the state of a whole DNA macromolecule, when recovery of all electronic ties on a sugar-phosphate framework, which were severed during the initial process of DNA degradation, takes place. The value T_1, determining probability $1/T_1$ by reduction of width of a double break determines average time of a reversible relaxation of a double break at the absence of potential barrier.

At the presence of anti-reparation barrier with height V_{max}, probability of reversible relaxation is determined by a product of two independent processes:

a) probability $exp(-V_{max}/k_b T)$ of one of the DNA fragments in a balanced environment with temperature T receiving enough energy by way of fluctuation for crossing the barrier towards decreasing a break;

b) probability $1/T_1$ considered above.

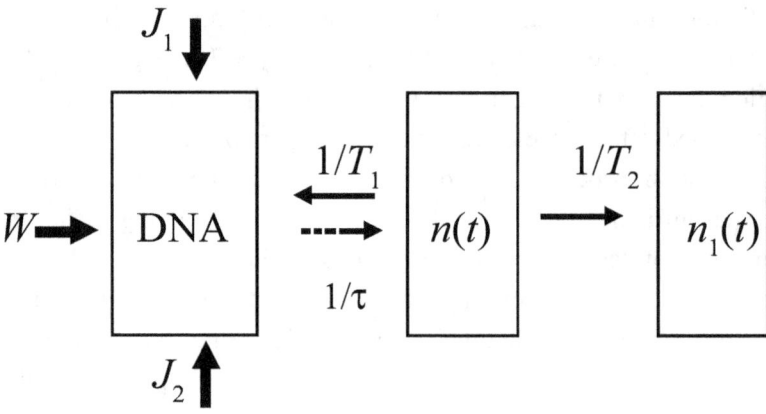

Fig. 4.2   Influence of two types of ionizing radiation (with respective intensities J_1 and J_2) independent of irradiation factor of degradation W on formation and relaxation of reversible n and irreversible n_1 double-string DNA breaks

The other part of relaxation processes leads to an outcome, when a break becomes irreversible. This is explained by an increase of the width of a double break L to such a value $L \gg L_0$, when elimination of that break becomes impossible. Probability of such a process at the absence of barrier V_{max} is determined according to general principles of thermodynamics, phenomenological value $1/T_2$. The value T_2, reciprocal to probability of relaxation, determines average time of irreversible relaxation of a double break.

Explicit relation between T_1 and T_2 may be determined from a full solution of a complicated statistical problem about fluctuational movement of DNA fragments. Based on the precise setting of the given problem, it is obvious that $T_1 \ll T_2$.

Should we account for probability of a reverse process – transition from a state of irreversible double break to a state of reversible break?

We think, that it would not be helpful.

This conclusion follows immediately from general principles of statistical physics and thermodynamics. The initial condition of a reversible double break corresponds to several specific locations of DNA fragments in the range $0 < L < L_0$ and in the area $L \approx L_0$. The number of those states corresponds to the final (limited by value) repetition factor of degeneration g_1 = const. Meanwhile, the state of irreversible double break includes all possible configurations of mutual locations of DNA fragments over the whole space $L > L_0$. The number of such locations is infinitely large. This situation is matched by an infinitely large repetition factor of degeneration of the state of irreversible double break and $g_2 \to \infty$. Meanwhile, it is well known, that probability of any directed process in statistical physics and thermodynamics is always determined by relation of partial probability of direct transition without taking the repetition factor of degeneration into account (in this case, the value of $1/T_2$ equal for any direction of transition) to corresponding repetition factor of degeneration of initial state of transition. Therefore, relation of probabilities of reverse and straight transitions is determined by relation of repetition factors of degeneration approaching zero $g_1/g_2 \to 0$. It follows directly from here that reverse transition from a state of irreversible break to a state of reversible break may be ignored. This circumstance makes further calculations significantly easier and it has a simple logical explanation. Indeed, any change of condition of a reversible state leads to only two possible outcomes – elimination of a break or formation of an irreversible break. At the other hand, any change of an irreversible break leads to an infinitely large number of variants – a DNA fragment may move in any direction and any distance. Only one type of movement (the case of a strictly reciprocal direction of severed fragments) leads to generation of a state of reversible break. Clearly, large selection of statistically equally likely processes makes this absolutely specific way of movement in a certain direction very unlikely.

Considering all the mentioned arguments, we shall rewrite the kinetic equation as

$$dn/dt = b_1 J_1 + b_2 J_2 + W - (n/T_1) \, exp\{-V_{max}/k_b T\} - n/T_2 \equiv$$
$$\equiv b_1 J_1 + b_2 J_2 + W - n/\tau \qquad (4.9)$$

describing the process of non-stationary dynamics of altering volume concentration $n(t)$ of reversible double DNA breaks and including dependence of changing $n(t)$ from parameters of environment, external ionizing fields with intensity $J_{1,2}$ and rate W of formation of double DNA breaks in a unit of volume under the influence of non-radiation factor of degradation of DNA (for example, the effect of free radicals).

The value

$$1/\tau = \{(1/T_1) \, [exp(-V_{max}/k_b T)] + 1/T_2\} \qquad (4.10)$$

means probability of sum (reversible and irreversible) relaxation of a double break.

If radiation is stopped at the moment of time Δt, solution of equation (4.9), determining concentration of reversible double DNA breaks during irradiation (i.e. at $t < \Delta t$) has the form

$$N(t < \Delta t) = (b_1 J_1 + b_2 J_2 + W)\tau[1 - exp(-t/\tau)] \qquad (4.11)$$

Respectively, a solution of equation (4.9), describing the dynamics of concentration of reversible double DNA breaks at $t > \Delta t$, i.e. after radiation is stopped, has the form

$$n(t \geq \Delta t) = n(\Delta t) \, exp[-(t - \Delta t)/\tau]$$
$$= (b_1 J_1 + b_2 J_2 + W)\tau[exp(\Delta t/\tau) - 1] \, exp(-t/\tau) \qquad (4.12)$$

It follows from (4.12) that, eventually, after radiation is stopped, a reversible concentration of double breaks is reduced to zero because some part of them self-repair (which causes DNA to regain its integrity), while another part is transferred to the state of irreversible breaks.

Changing concentration n_1 of irreversible double breaks, accounting for a possibility of ignoring reverse transition of an irreversible double DNA break, mentioned earlier, back to the state of reversible break, is described by equation

$$dn_1/dt = (n/T_2) \qquad (4.13)$$

Using solutions for equations (4.11) and (4.12), obtained earlier, for concentration of reversible breaks $n(t)$, we find an expression for $n_1(t)$ at $t < \Delta t$

$$n_1(t < \Delta t) = (b_1 J_1 + b_2 J_2 + W)\tau\{-[1 - exp(-t/\tau)](\tau/T_2) + t/T_2\} \qquad (4.14)$$

Correspondingly, at $t \geq \Delta t$, from (4.13) we find

$$n_1(t \geq \Delta t) = (b_1 J_1 + b_2 J_2 + W)(\tau/T_2)\{\Delta t - \tau \ exp(-t/\tau)[exp(\Delta t/t) - 1]\} \qquad (4.15)$$

Resulting concentration of irreversible double breaks follows from (4.15) at infinitely increasing time t after radiation is stopped

$$n_1(t \rightarrow \infty) = (b_1 J_1 + b_2 J_2 + W)\tau \Delta t/T_2 \qquad (4.16)$$

In these equations:

$B_{(1,2)} = 2n_d N_n \sigma_{(1,2)d}\beta_{(1,2)d} + n_3\sigma_{(1,2)3}\beta_{(1,2)3}$ – parameters determining relation between intensity of ionizing irradiation J_1 and J_2 with concentration n of reversible double breaks, induced by them;

n_d – concentration of DNA macro molecules in inner cellular environment; $2N_n$ – number of atoms in sugar-phosphate frame of DNA;

σ_{1d} and σ_{2d} – average cross section areas of absorption of two types of ionizing radiation being considered by atoms of the sugar-phosphate DNA frame;

$1/\beta_{1d}$ and $1/\beta_{2d}$ – average energy of radiation fields J_1 and J_2, absorbed during ionization of atoms in sugar-phosphate DNA frame, per one double break (these values determine efficiency of the process, resulting from impact of direct radiation); $1/\beta_{13}$ and $1/\beta_{23}$ – average energy of radiation fields J_1 and J_2, absorbed during hydrolysis, per one double break (these values determine efficiency of the process, resulting from impact of indirect radiation);

W – rate of formation of double breaks in a unit of volume of a liquid per unit of time stipulated for by mechanism of DNA degradation due to an effect of free radicals of non-radiation origin.

Relation of variables β_{13}, β_{23}, β_{1d} and β_{2d} to average energy $E_{1p,2p} = 1/\beta_{1p,2p}$, corresponding to a double DNA break, is defined by the following expressions

$$\beta_{p(1,2)} = (2n_d N_n \sigma_{(1,2)d}\beta_{(1,2)d} + n_3\sigma_{(1,2)3}\beta_{(1,2)3})/(2n_d N_n \sigma_{(1,2)d}\beta_{(1,2)d} + n_3\sigma_{(1,2)3}) \approx$$
$$\approx [\beta_{(1,2)d}(2n_d N_n/n_3)(\sigma_{(1,2)d}/\sigma_{(1,2)3}) + \beta_{(1,2)3}]$$

accounting for the natural condition of low concentration of atoms in sugar-phosphate DNA frame in comparison with concentration of water molecules, which corresponds to the relation $n_d N_n \ll n_3\sigma_{(1,2)3}/2\sigma_{(1,2)d}$.

4.1.2   Activated water and stability of DNA macromolecules

Solutions (4.11) – (4.16) for kinetic equations (4.9) and (4.13) obtained above determine reaction of a biological system on the impact of various factors of degradation of DNA. They characterize the process of evolution of damaged DNA macromolecules in natural water-salt inner cellular medium.

What would change in these equations and their solutions, if inner cellular water were activated?

As we have shown in chapter 2, the process of water activation is accompanied by a significant change of its structure and parameters. This change may be manifested on two different hierarchical levels of water structure organization.

One of these levels (the macro level of water structure) corresponds to the spatial water structure and determines the shape and structure of its spatial frame. Keep in mind, that the macroscopic level of water structure corresponds to the system of clathrate hydrates, which generate stable dodecahedric polyhedrons from ions of oxygen and hydrogen. Inside these polyhedrons there are micro cavities. Polyhedrons are connected by hydrogen links into associates. Under certain conditions, these associates may be combined into larger associates (macro clusters). The space between the associates and macro clusters is filled with quasi amorphous water.

The second hierarchical level (micro level of water structure) characterizes special features of filling of micro cavities of the spatial clathrate frame of water with separate H_2O molecules. This micro level determines evolution of H_2O molecules, which can leave the volume of quasi amorphous water, penetrate inside those micro cavities and stay there in hydrophobic form for a long time, or, to the opposite, transfer from micro cavities into the volume of quasi amorphous water.

Apparently, the micro level of water structure is distinguished by a better stability to external destructing factors than the macro level. With all external transformations of the clathrate frame, hydrophobic molecules H_2O remain in stable condition in the volume of micro cavities.

Let's investigate the micro level of water structure in greater detail.

During water activation the ratio of the number of H_2O molecules contained in "regular" (or quasi amorphous) water to the number of molecules contained in isolated micro cavities in the structure of clathrate hydrates is altered. Such redistribution, given the specifics of biological objects, has an obvious explanation. High thermal stability of a human organism means that all water contained in it should have a strictly fixed number of micro cavities filled with H_2O molecules, corresponding to normal temperature, as well as a certain concentration of molecules of quasi amorphous water (*Table 2.1*). This relation is a result of the Bolzman distribution for a balanced system in regular (non-activated) water. Throughout the entire human evolution, this order remains intact.

Another situation represents activated unbalanced water. During its activation with the help of preliminary external influence, unbalanced distribution of a population of H_2O molecules in micro cavities is achieved. It may be significantly different from its value, corresponding to thermodynamic balance at normal temperatures. Depending on the type of water activation, various deviations of population distribution in either direction are possible. If we compare unbalanced population distribution with standard balanced distribution in the same system at a given actual temperature T_{eff}, condition of activated water may be characterized by its actual (effective) temperature. Clearly, depending on method of activation, actual (effective) temperature of water T_{eff} may be higher or lower than average water temperature T, which is a temperature of its quasi amorphous part.

If a human organism receives activated water, parameters of water environment change proportionately to its concentration in various organs. From results of calculations conducted in chapter 2 we see that at a normal body temperature, water can remain in activated state for quite a long time (no less than 24 hours), which would be sufficient for a therapeutic effect such water on human organism.

This water has other physical and molecular features.

If, during activation, "extra" H_2O molecules are found in the volume of micro cavities in clathrate hydrates, while in quasi amorphous water there is a deficit of such molecules, common concentration and viscosity of such water will increase. These changes of viscosity and density are corroborated by numerous experimental results on water magnetization by a direct magnetic field. In the simplest case, a changed distribution is similar to results obtained during thermal activation during a rapid cooling of hot water to its physiological norm. In this case, the average temperature of water corresponds to such a norm, while distribution of molecules in the volume of micro cavities corresponds to the initial high temperature.

In the opposite case of water molecules' deficit in the volume of micro cavities and their excess quantity in quasi-amorphous water, total density and viscosity of water decrease. In case of a similar thermal activation, such distribution may be obtained during a very fast heating of water from melted snow to the same physiological temperature. In this case, effective temperature T_{eff}, determining distribution of water molecules in micro cavities, will be low.

Evidently, the maximal change of relation between isolated and quasi-free water molecules is limited by interval of temperature, used in activation (not lower than 0°C and not higher than 100°C). Other methods of water activation (for example, with the help of periodically changing magnetic field) are free of this limitation and, therefore, they, supposedly, may be used in achieving a state of water with an even higher degree of activation.

Alternating magnetic field influences water, which is a diamagnetic. Its magnetic permittivity is less than one ($\mu < 1$). Volume density of the forces, acting on a unit of volume of water, equals

$$f_M = (\mu - 1)\nabla H^2/8\pi = -(1 - \mu)\nabla H^2/8\pi$$

We can see, that at such a diamagnetic, at $\mu < 1$, force of magnetic pressure will be directed against the gradient of an alternating magnetic field. This force will try to change distribution of water molecules in such a way, that they are transferred to an area of weaker magnetic field. In a periodically alternating field, the direction and force of magnetic pressure will also be alternating. That causes periodical and correlated collision of clathrate frames of separate clusters. Such collisions may with higher probability stimulate formation of quasi stable macro clusters in a volume of water, each one of which consists of a system of reciprocally ordered clathrate frames.

These macro clusters change the structure of water on the macro level. Ordered macro clusters are principally different from short-living "flickering clusters" generated during random and uncorrelated collisions of clathrate frames, which occurs during random thermal movement of separate clusters. First of all, due to a better-ordered nature of collisions, macro clusters will have more stable hydrogen links, which causes a generally higher stability. Presence of macro clusters significantly changes viscosity and other mechanical characteristics of water, which inevitably affects the process of DNA reparation.

Another difference constitutes the fact that in ordered structures of macro clusters there will be periodically placed micro cavities. In such system, due to transmission symmetry, energy zones of allowed states are generated. Within these areas, water molecules will move freely between micro cavities, which may lead to rapid activation of the clathrate frame of water.

Periodical features may affect not only movement of the clathrate frame, but alter its features as well, such as, for example, transparency of a potential barrier in windows of micro cavities. This task corresponds to possibility of faster tunneling of H_2O molecules through non-stationary potential barrier.

Another aspect of influence of a strong periodical magnetic field is its ability to stimulate transitions between energy levels, which characterize condition of H_2O molecules in micro cavities and quasi-amorphous water due to multiple-photon non-linear processes during interaction with magnetic moments. Such non-linear processes may also lead to fast activation of water and establishing of unequal population distribution of water molecules in micro cavities.

With the help of such coherent effects, it is possible to create such an unequal population distribution of H_2O molecules in micro cavities, which would be impossible to achieve during a temperature change.

It is interesting to note, that an electric field (both constant and alternating) will also influence properties of water, but in this case the main mechanism of such influence will work through affecting quasi-amorphous water and related to polarization of dipole moments of H_2O molecules.

There are other, no less important, aspects of activated water's influence on physiological processes and DNA stability.

Since water molecules in the volume of micro cavities in clathrate frame have hydrophobic properties and don't form hydrogen links with surfaces of micro cavities, such molecules have altered dispersion characteristics in comparison with thermodynamically balanced water at the same temperature. Such water will have a different coefficient of radiation absorption over the whole specter from microwave to ultraviolet frequency ranges, a different refraction coefficient, different values for actual $\varepsilon'(\omega)$ and imaginary $\varepsilon''(\omega)$ parts of dielectric permittivity. Naturally, in such activated water the ratio of Coulomb forces of interaction to forces of dispersion interaction of the Van Der Waals type in the area of a double break of DNA macro molecules would also be different. Importance of this circumstance becomes evident if we recall that the balance of these forces determines availability and parameters of the anti-reparation barrier, which hinders the process of self-reparation of DNA macro molecules.

It was noted earlier, that a deviation of the current distribution of H_2O molecules in micro cavities of clathrate hydrates and the volume of a quasi amorphous water with weak structure from a balanced state at a given temperature may be described by an actual (effective) temperature T_{eff}.

Is it permissible to claim that an effect of activated unbalanced water on DNA macro molecules in human organism at its natural physiological temperature of 36.6°C is equivalent to the effect of balanced water taken at that equivalent temperature? It's easy to verify that such comparison would be wrong.

As a matter of fact, the process of interaction of DNA fragments in the volume of inner cellular media accounts not only for dispersions of water media and interacting pairs of nucleotides, but the ionic content of a water media, which, itself, strongly depends on temperature. Both these factors are quite equal in their relation to influence on the process of DNA self-reparation. And if, in activated water, distribution of H_2O molecules, regulating its dispersion, actually corresponds to some equivalent temperature (which at a large deviation from the state of thermodynamic equilibrium may be located even outside the interval of liquid phase existence) that means that quasi-amorphous water is characterized by real physiological temperature of a living organism.

Unfortunately, at the present time, dispersion characteristics of activated water in various diapasons of the electromagnetic specter are practically unknown. One exception is a study of water refraction coefficient's dependency on temperature in an optical diapason and in constant electrostatic field. For example, based on research results (Zenin, 1994), coefficient of refraction in the optical diapason for water at the presence of a spatial structure in the form of clathrates (such structure corresponds to low temperature of water) is $n_c = 1.338$, while for regular quasi amorphous water (at a high temperature) it is $n_0 = 1.286$. These results mean, in fact, a study of a macro level of activated water (it was noted before, that such level signifies the presence of an ordered clustered structure of water). Unfortunately, these results are not helpful in obtaining information whether cooled water was activated on the micro level, or it was in the state of thermodynamic equilibrium with regards to distribution of H_2O molecules between micro cavities and the volume of quasi amorphous water.

The main parameter determining the sign and value of the force of Van Der Waals dispersion interaction is dielectric permittivity $\varepsilon_w(\omega) = n(\omega)^2$. Relative difference of refraction coefficients for regular and structured water is about 5%, which corresponds to a change of relative dielectric permittivity by 10% and can significantly alter the nature of DNA self-reparation process. There are good reasons to believe, that with additional activation of water on the micro level, changes of dielectric permittivity would be even more significant. If similar changes happen in other areas of the specter (primarily, the UV range), it may result in an additional increase of dielectric permittivity. According to general patterns discussed in chapter 3, an increase of dielectric permittivity would cause an intensification of Van Der Waals dispersion interaction (its energy should increase), which would result in reduction of the height of potential barrier in the area between severed ends of DNA.

4.2   FEATURES OF DEGRADATION OF A SYSTEM OF DNA MACRO MOLECULES AT COMBINED IONIZING IMPACT

4.2.1   Relation of degree of DNA degradation with duration and intensity of ionizing impact

Based on solutions (4.11)–(4.16) we shall conduct an analysis of reaction of a system of DNA macro molecules on different types of combined ionizing influence, accounting for differences in their intensities, doses, efficiency, duration, as well as properties of the media. We shall consider special features of DNA degradation in time under the influence of ionizing fields and other factors of degradation of DNA.

From (4.11) and (4.14) we find full concentration of all (reversible and irreversible) double breaks of DNA during the effect of radiation (i.e. at $t < \Delta t$)

$$N(t < \Delta t) = n(t < \Delta t) + n_1(t < \Delta t) =$$
$$(b_1 J_1 + b_2 J_2 + W)\tau\{[1 - exp(-t/\tau)](1 - (\tau/T_2)) + t/T_2\} \qquad (4.17)$$

This variable simultaneously determines the efficiency of destructive effect on DNA and efficiency of the system of DNA self-reparation. Proceeding from analysis of (4.17), we can investigate patterns of DNA evolution depending on specific parameters of various types of ionizing influence.

a)   Influence of ionizing radiation with random intensity and very short duration on DNA

We shall begin investigation of this problem with analysis of the case of short-duration impact of ionizing factors on DNA (ultraviolet, Roentgen or gamma-radiation and free radicals). If duration of radiation Δt is less than all times of relaxation ($\Delta t \ll \tau, T_1, T_2$) then, from solutions (4.11), (4.14) and (4.17) we directly find

$$N(t) \approx n(t \leq \Delta t) \approx (b_1 J_1 + b_2 J_2 + W)t = b_1 Q_1 + b_2 Q_2 + Wt \qquad (4.18)$$

$$n_1(t) \approx (b_1 J_1 + b_2 J_2 + W)t^2/2T_2 \ll n(t \leq \Delta t)$$

It can be seen that in this case full volume concentration N of all double breaks changes linearly, in direct proportion to obtained densities of doses $Q_1 = J_1 t$, $Q_2 = J_2 t$, Wt.

This result complies with rules of additivity and linearity of the effects of both types of radiation and non-radiation factors of degradation, regardless of intensity of radiation and amount of exposure: the result of the combined effect of each of the factors of DNA degradation is equal to the sum of results of effects of the same factors.

This result, representing the case of an extremely short-duration radiation fully complies with traditional radiobiological concepts of universality of the rule of additivity during combined irradiation. In this respect it is necessary to note that in applied radiobiology, the rule of additivity is usually generalized (without adequate justification) in order to account for cases of irradiation with random duration. We shall demonstrate later why such assumption is wrong.

b)   Influence of ionizing radiation of random intensity and medium duration on DNA

When DNA is subjected to the effect of ionizing radiation with medium duration $T_2 \gg \Delta t \gg T_1$ and random intensity, from (4.17) we have

$$N(t) = n(t) + n_1(t) = (b_1 J_1 + b_2 J_2 + W)T_1(1 + t/T_2) \approx (b_1 J_1 + b_2 J_2 + W)T_1 \quad (4.19)$$

Note, that in this case there is also the effect of additivity of action of both types of radiation and non-radiation factors of degradation. There is also a significant difference from the previous case of very short-duration irradiation – concentration of double breaks N depends on intensities J_1, J_2, W of radiation fields and non-radiation factor, but it doesn't depend on exposure to radiation. For radiation with such intermediate duration, harmful effect of radiation fields and non-radiation factors decreases in comparison with forecasted linear law of additivity by $t/T_1 \gg 1$ times.

c)   Influence of random doses at long time of exposure of DNA

With long exposure times (at $\Delta t \gg \tau, T_1, T_2$) and with large total dose of radiation, from (4.17) we find

$$N(t) \approx (b_1 Q_1 + b_2 Q_2 + Wt)(T_1/T_2) \, exp(V_{max}/k_b T) \qquad (4.20)$$

We shall consider this result more thoroughly.

For relatively weak radiation fields J_1 and J_2, values of which are limited by the condition

$$[(\beta_{1e} + \beta_1^*)\sigma_{13} J_1 + (\beta_{2e} + \beta_2^*)\sigma_{23} J_2]\tau_e \leq 1/G n_3 \qquad (4.21)$$

a linear approximation (4.1) is acceptable.

Given (4.1), from equation (4.17) we find

$$N(t) \approx (b_1 J_1 + b_2 J_2 + W)t(T_1/T_2) \, exp[V_0(1 - J_1/J_{01} - J_2/J_{02})/k_b T] \equiv$$

$$\equiv (b_1 Q_1 + b_2 Q_2 + Wt)(T_1/T_2) \, exp[V_0(1 - Q_1/J_{01}t - Q_2/J_{02}t)/k_b T] \qquad (4.22)$$

Here $J_{0(1,2)} = 1/[G n_3 \tau_e (\beta_{(1,2)e} + \beta_{(1,2)}^*)\sigma_{(1,2)3}]$ – characteristic effective intensities determined by parameters of inner cellular water-salt media and depending on special features of interaction of atoms and molecules of that media with specific types of ionizing radiation.

It follows directly from (4.22) that the dependency of concentration of double DNA breaks from radiation intensity has non-linear nature. Meanwhile, values of effective intensities J_{01}, J_{02}, which depend on characteristics of interaction of two types of radiation with nucleotides and inner cellular media, have the meaning of extreme intensities, separating areas of linear and non-linear degradation of DNA macro molecules.

Let's analyze (4.22) in its relation from radiation intensities.

In case of large doses of radiation, complying with the condition

$$Q_1/J_{01}t + Q_2/J_{02}t \geq 1,$$

from (4.22) we have

$$N(t) \approx (b_1 Q_1 + b_2 Q_2 + Wt)(T_1/T_2) \qquad (4.23)$$

Since $T_1/T_2 \ll 1$, we can see, that the case of a large dose of radiation corresponds to traditional concepts of radiobiology – reduction of the number of double breaks by $(T_2/T_1) > 1$ times and, consecutively, the degree of DNA degradation for a long time of radiation t (compared with the case of short-duration irradiation (4.18)).

In other words, for the case of large doses of radiation, the maximum DNA degradation at a constant dose corresponds to short-duration ("sharp") irradiation.

We would like to remind, that the presence of a potential barrier in the space between separated ends in the area of a double DNA break is determined by balance of Coulomb forces of electrostatic interaction and Van Der Waals forces of dispersion interaction. In normal inner cellular liquid environment, the combined result of those interactions leads too appearance of a potential barrier with height V_{max} between oriented pairs of nucleotides *GC–GC* and *AT–AT*. In activated water parameters of interaction change, which causes deformation of the barrier and reduction of its height.

From generated results (4.18), (4.19) and (4.23) we can see, that results of influence by large doses (at any duration of exposure) do not directly depend on the presence of a potential barrier and its height V_{max}. That means that in case of radiation of random intensity and small (4.18) or medium (4.19) duration, and in case of (4.23) of radiation by a large dose, the result of ionizing influence on DNA does not depend on whether water is activated or not.

Accordingly, for another extreme case (small doses of ionizing radiation Q_1 and Q_2, complying with the condition $Q_1/J_{01}t + Q_2/J_{02}t \ll 1$) the result (4.22) is reduced to a linear equation

$$N(t) \approx (b_1 Q_1 + b_2 Q_2 + Wt)(T_1/T_2)\, exp(V_0/k_b T) \qquad (4.24)$$

In this case the degree of DNA degradation is differs from the case of influence on DNA of ionizing radiation of an extremely short duration (4.18) by a non-dimensional parameter $(T_1/T_2)\, exp(V_0/k_b T)$. The value of that parameter depends on specific relation between times of reversible and irreversible relaxation $(T_1/T_2) < 1$ and probability of thermally stimulated transformation of a particle above the barrier $exp(V_0/k_b T) > 1$.

The general form of this functional dependence is illustrated on *Fig. 4.3*.

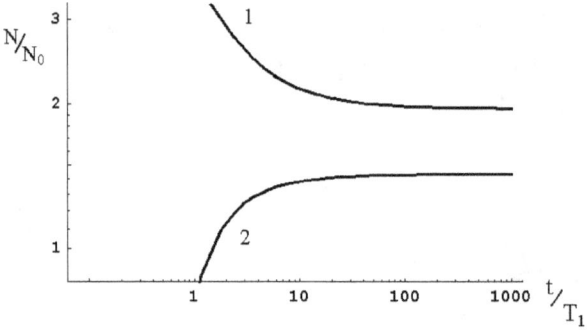

Fig. 4.3 *Features of influence of small and large doses during short-duration (sharp) and long (chronic) irradiation of DNA. 1 – irradiation with a large dose $Q_1 = 10\, J_{01}T_2$; 2 – irradiation of DNA with a small dose $Q_2 = 0.1\, J_{02}T_2$. Calculation corresponds to a double DNA break with identical end pairs "guanine-cytosine" and relation between times of irreversible and reversible relaxation $T_2/T_1 = 10$*

It can be seen that during irradiation with a small dose with the presence of a sufficiently high barrier V_0 (for which condition $exp(V_0/k_b T) > T_2/T_1$ is satisfied) there is an opposite (in relation to the effect, considered earlier and described by (4.23)) and quite paradoxical effect – acceleration of DNA degradation at the same full doses and extension of exposure time compared with the case of short-duration irradiation (4.18). That result tells us that for the case of small doses, the highest degree of DNA degradation corresponds to long ("chronic") irradiation. This surprising result (inversion of the effect of short-duration and long-duration irradiation) was detected recently in experimental radiobiology in detailed analysis of the results of influence by moderately small doses. Note, that the amount of moderately small doses is considerably higher than extremely small doses, considered later, causing the effect of hormesis.

The reason for this amazing result is obvious and follows immediately from a direct comparison of the initial expression (4.22) with expressions (4.23) and (4.24).

The first part of these formulae (first efficient $(b_1Q_1 + b_2Q_2 + Wt)$) is proportionate to the sum dose of ionizing influence and, naturally, increases with growing dose, but doesn't depend on exposure time.

The second part of (4.22) is proportionate to the value

$$exp[V_0(1 - Q_1/J_{01}t - Q_2/J_{02}t)/k_bT]$$

and characterizes the process of reverse reparation of a double DNA break, depends on the height of a potential barrier $V_{max} = V_0(1 - Q_1/J_{01}t - Q_2/J_{02}t)$, which determines the probability of such reparation.

With small duration of irradiation (and the same unchangeable total dose), the height of a potential barrier is reduced due to an effective action of hydrated electrons, for example, which makes the process of self-reparation easier. Meanwhile, in case of an extended impact by the same dose, concentration of hydrated electrons is small, while the barrier is hardly reduced, which makes the process of self-reparation considerably harder and leads to an increase of concentration of damaged macro molecules of DNA.

On *Fig. 4.4* there is a detailed demonstration of specifics of influence of small $(Q << J_0T_2)$, intermediate $(Q \approx J_0T_2)$ and large $(Q >> J_0T_2)$ doses of ionizing radiation on the entire quantity N of double DNA breaks with a short-duration ("sharp", $t << T_1$) and long-duration ("chronic", $t >> T_1$) radiation of DNA as well as varying ratio of times of reversible and irreversible relaxation. These functional dependencies demonstrate all the mentioned patterns including inversion of impact of sharp and chronic radiation.

On *Fig. 4.5* there is relation between changes of nominal (for parameter b) concentration of irreversible double DNA breaks and the height of a potential barrier and the dose of ionizing radiation, calculated on the basis of obtained results.

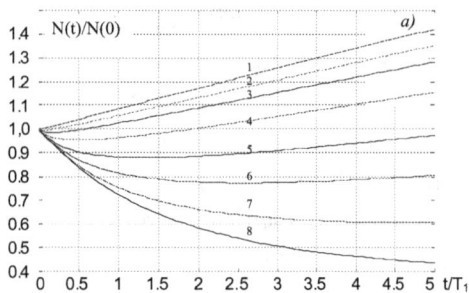

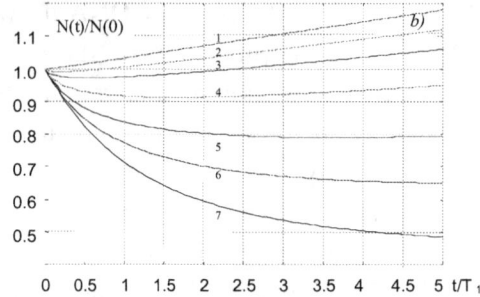

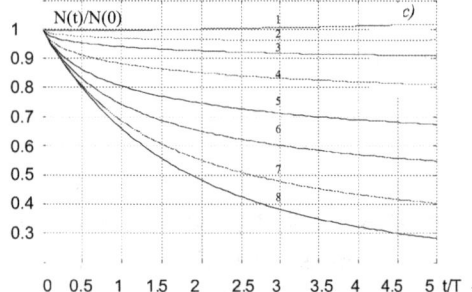

Fig. 4.4   Features of influence of small and large doses of ionizing radiation on the entire quantity N of double DNA breaks during short-duration ("sharp", t << T₁) and long-duration ("chronic", t >> T₁) irradiation of DNA. 1 – irradiation of DNA with a dose of $q = Q/J_0T_2 = 0.01$; 2 – q = 0.1; 3 – q = 0.3; 4 – q = 1; 5 – q = 3; 6 – q = 7; 7 – q = 20; 8 – q = 100. This calculation relates to a double DNA break with identical end pairs "guanine-cytosine" and the ratio of times of irreversible and reversible relaxation: a) $T_2/T_1 = 4$; b) $T_2/T_1 = 7$; c) $T_2/T_1 = 14$; d) $T_2/T_1 = 50$.

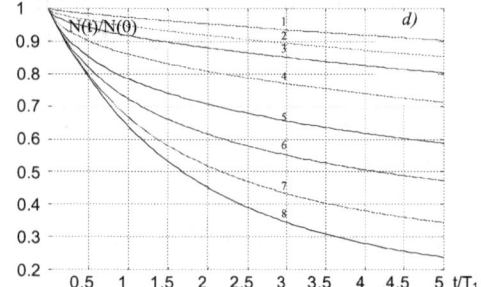

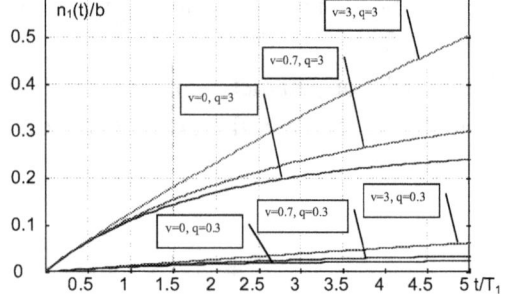

Fig. 4.5   Functional dependence of changes of normative (for parameter b) concentration of irreversible double DNA breaks from relative height of anti-reparation potential barrier $V = V_0/k_bT$ and normative dose of ionizing radiation $q = Q/J_0T_2$.

4.2.2. The "Hormesis" effect and the mechanism of self-induced protection from radiation and influence of free radicals in regular and activated water

Using (4.17), we shall conduct an analysis of specific features of combined simultaneous effect of several types of ionizing radiation and non-radiation factor of DNA degradation on DNA. In this section we shall study the mechanisms of reciprocal influence of various factors on the general problem of DNA stability. The result of such reciprocal influence is effect of radiation antagonism and "hormesis" effect. In order to study these effects in more detail, let's consider reciprocal effect of various factors of influence.

a) Specific features of reciprocal influence of ionizing radiation and the factor of ionizing DNA degradation with long period of irradiation

We shall consider specific features of the dynamics of creation and reparation of double DNA breaks under the impact of only one type of ionizing radiation with intensity J. In this case, from (4.20), including functional dependence (4.1) and in full analogy with (4.22), we have

$$N(t) \approx (bJt + Wt)(T_1/T_2) \, exp[V_0(1 - J/J_0)/k_bT]$$

The latter expression may be represented explicitly for three characteristic areas of dependence of concentration N of double DNA breaks from intensity J.

a1) With low intensity of external irradiation $J \ll J_0$, concentration of double breaks $N(t)$ increases linearly, in proportion to both time of irradiation t and intensity of ionizing radiation J

$$N(t) \approx (bJt + Wt)(T_1/T_2) \, exp(V_0/k_bT) \qquad (4.25)$$

a2) With intermediate intensity $J \leq J_0$, there is a non-linear dependence of concentration $N(t)$ from intensity of ionizing radiation J

$$N(t) \approx (bJt + Wt)(T_1/T_2) \, exp(V_0(1 - J/J_0)/k_bT) \qquad (4.26)$$

From (4.26), with a certain combination of parameters of system $J_0 < WV_0/bk_bT$, depending on specific efficiency of impact of free radicals W, it means one very particular and amazing phenomenon – reduction of DNA degradation (i.e. increase of its resistance to radiation) with increasing of intensity of ionizing irradiation J.

This effect is described by equation

$$N(t) \approx \{W - J[WV_0/J_0k_bT) - b]\}t(T_1/T_2) \, exp(V_0/k_bT), \qquad (4.27)$$

which follows from (4.26) provided that condition $J < J_0 < WV_0/J_0 k_b T$ is satisfied and represents the "Hormesis" phenomenon – positive influence of an extremely low-dose chronic irradiation with intensity $J < J_0$ on biological system.

From (4.27) we can see that these conditions correspond to a paradoxical situation – reduction of concentration of double DNA breaks with increasing intensity of radiation J (*Fig. 4.6*).

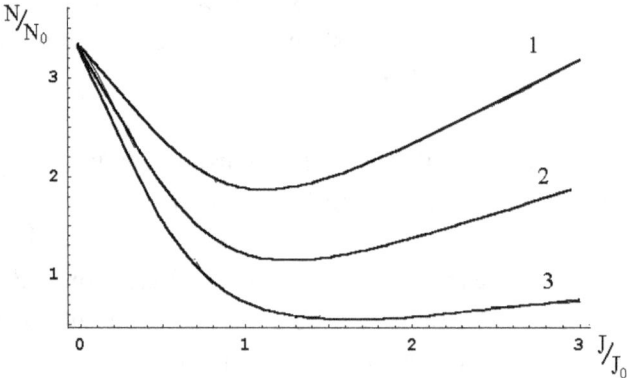

Fig 4.6   *The "Hormesis" effect with simultaneous effect of ionizing radiation of low intensity J and factor of degradation W, independent of radiation, on DNA. Calculation of the "Hormesis" effect corresponds to a double DNA break with identical end pairs "guanine-citosine" (natural height of the barrier $V_0 = 3k_b T$) and the ratio of irreversible and reversible times of relaxation $T_2/T_1 = 10$. Variable $N_0 \int N(J = 0)$ represents DNA degradation in the absence of radiation.*

Dependencies 1-3 are determined with values $J_0 b = 0.1; 0.5; 1$.

Simple analysis of (4.26) shows that function $N(t)$ decreases to its minimal value

$$N_{min} \approx (bJ_0 + W)t(T_1/T_2)$$

When intensity of irradiation J reaches its optimal value $J^{opt} \equiv J_0$.

Within the framework of the studied model, the cause of the "hormesis" effect is obvious. It's easy to understand, if we recall, that when ionizing radiation affects a DNA system, two reciprocal effects occur.

On one side, under the influence of that radiation new DNA breaks are created, but, on the other side, potential barrier regulating the process of self-reparation of double DNA break is deformed. For that reason, it's obvious, that the reason for the "hormesis" effect lies in the fact that when conditions $J < J_0$, $J_0 < WV_0/J_0 k_b T$, mentioned earlier, are satisfied, ionizing irradiation J has a weak effect on the possibility of formation of new double breaks, while reduction of the height of self-reparation barrier V_{max} is very well defined. That reduction, in its turn, contributes to self-elimination of those double breaks, which were caused by processes, unrelated to irradiation (in this case, effect on free radicals of non-radiation origin on DNA).

The same result may be formulated more clearly in the following way.

The essence of the process of "hormesis" is the fact that an extremely weak ionizing radiation doesn't cause additional DNA damage, but creates conditions for elimination of damage to DNA, caused by the effect of free radicals of non-radiation origin.

When the opposite condition $J_0 > WV_0/bk_bT$ is fulfilled, the effect of "hormesis" is absent, but there is the process of relative suppression of processes of radiation-caused destruction of DNA (self-antagonism of radiation) by the value $exp(-V_0J/J_0k_bT)$. Maximum relative suppression of DNA degradation corresponds to optimal intensity $J^{opt} \equiv J_0$. "Hormesis" stops when efficiency of generation of new breaks exceeds that of the mechanism, stimulating the process of DNA self-reparation.

a3) With high intensity of external irradiation $J > J_0$ there is a linear increase N with growth J, independent of relationship between W, J_0 and b, described by the formula

$$N(t) \approx (bJ + W)t(T_1/T_2) \tag{4.28}$$

We can see that in the case of large intensity, there is linear increase of total concentration of double breaks from a dose of ionizing radiation and effective dose of action of free radicals.

a) Particular qualities of reciprocal influence of several types of ionizing radiation on DNA and conditions for existence of radiation antagonism.

We shall consider special features of another non-linear effect – combined influence on DNA of two types of radiation J_1 and J_2, distinguishable by their ability to produce double breaks (different parameters b_1, b_2) and affect the anti-reparation barrier (different parameters c_1J_{01}, c_2J_{02}). With simultaneous influence of radiation of one type with low intensity $J_1 \ll J_1^{opt} \equiv J_{01}$ and another type of radiation with random intensity J_2 on DNA, dependence of concentration of double breaks is determined by (4.22), which determines possibility of radiation antagonism. Let's consider conditions and specific features of such effect.

At the presence of one type of radiation with intensity, which is much lower than threshold intensity ($J_1 \ll J_{10}$) and absence of another radiation ($J_2 = 0$), from (4.22) follows the situation investigated above – linear increase of concentration of breaks from intensity of a field J_1 and time of exposure (i.e. dose Q_1).

$$N(t) = (b_1J_1 + W)t(T_1/T_2) \, exp[V_0(1 - J_1/J_{01})/k_bT] \approx (b_1J_1 + W)t(T_1/T_2) \, exp(V_0/k_bT)$$

With an additional effect of another weak radiation with intensity $J_2 \ll k_bTJ_{02}/V_0$, from (4.22) we have

$$N(t) \approx [b_1J_1 - J_2(b_1J_1 - b_2V_0/k_bT) + W] \, t \, (T_1/T_2) \, exp(V_0/k_bT) \tag{4.29}$$

From this expression we can see that if for intensity J_1, along with condition $J_1 \ll J_{10}$, another condition $J_1 > b_2V_0/b_1k_bT$ is also satisfied, influence of additional radiation with intensity J_2 will lead to another unusual result – absolute decrease of the number of double breaks and, correspondingly, reduction of degree of DNA degradation with increasing intensity J_2 (*Fig. 4.7*).

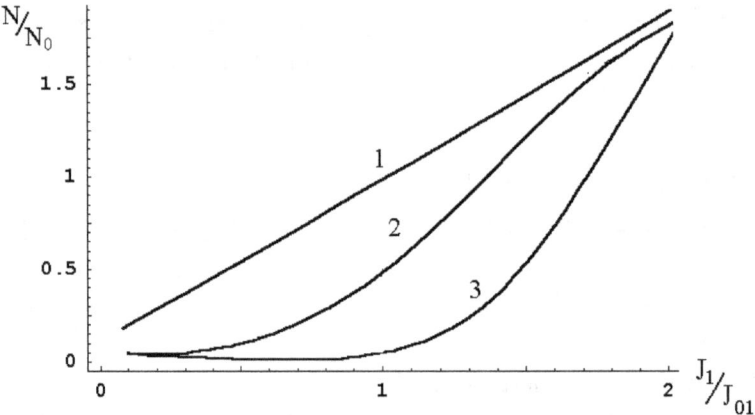

Fig. 4.7 Phenomenon of radiation antagonism at combined effect of two types of ionizing radiation on DNA. Calculation corresponds to the height of barrier $V_0 = 3k_bT$, the ratio of times of relaxation $T_2/T_1 = 10$ and parameters of media $(J_{01}b_1)/(J_{02}b_2) = 10$.
Dependencies 1-3 correspond to values $J_2/J_{02} = 0; 0.5; 0.9$.

This situation relates to another amazing phenomenon in radiobiology – *radiation antagonism*, when one type of radiation (characterized by its individual parameters in relation to generation of new DNA breaks and its influence on anti-reparation barrier) is described by explicitly defined protective functions in relation to another type of radiation (with its own individual parameters). Such reduction will occur at increasing J_2, up to an extreme (optimal) value

$$J_2^{opt} = J_{02}(1 - J_1/J_{01}), \tag{4.30}$$

with corresponding minimal concentration of double breaks (minimal degradation) of DNA

$$N_{min}(t) \approx [J_1(b_1 - b_2J_{02}/J_{01}) + b_2J_{02} + W] \, t \, (T_1/T_2) \tag{4.31}$$

Variable $N_{min}(t)$ is less than $N(t)$ under the influence of only one radiation J_1 by K times:

$$K = \{n(t) + n_1(t)\}/\{n(t) + n_1(t)\}_{min} \approx b_1J_1 \, exp(V_0/k_bT)/[J_1(b_1 - b_2J_{02}/J_{01}) + b_2J_{02}] \tag{4.32}$$

Only with $J_2 > J_2^{opt}$ an increase of total concentration $N(t)$ of double DNA breaks begin to happen.

At critical intensity of radiation J_2^{cr}, determined from condition

$$(b_1 J_1 + b_2 J_2^{cr}) = b_1 J_1 \, exp[V_0(1 + J_2^{cr}/J_{02})/k_b T] \qquad (4.33)$$

concentration of double breaks $N(t)$, generated due to influence of one type of radiation, with intensity J_1 equals analogical concentration $N(t)$ with combined influence of both fields J_1 and J_2^{cr}.

For $J_2 > J_2^{cr}$, total concentration of double breaks at the presence of J_1 and J_2 exceeds a value of N(t), which would be caused by an impact of only one type of radiation J_1.

If intensity of the first type of radiation is small ($J_1 < J_1^{opt}$, $J_1 < b_2 V_0/b_1 k_b T$), inclusion of a second type of radiation J_2 immediately leads to an increase of N, which cannot be compensated by self-reparation.

Explanation of these effects is quite simple. They are based on relations of the balance between two mutually opposite processes, caused by ionizing radiation – generation of additional breaks and influence on mechanisms of their self-reparation. When one radiation is intense $J_1 > b_2 V_0/b_1 k_b T$, influence of another radiation J_2 has a weak effect on generation of double breaks, but it has a very strong effect on appearance of additional hydrated electrons and agitated atoms, which provides efficient functioning of the mechanism of self-reparation. This case is characterized by minimal values of parameters b_2 and J_{02}. These requirements optimize the value of minimal $J_1 > b_2 V_0/b_1 k_b T$, from which the investigated radio-protective mechanism works, as well as the value of an additional field $J_2^{opt} = J_2(1 - J_1/J_{01})$. This process of reparation reaches its maximum efficiency at $J_2 = J_2^{opt}$ and then increasing with growth of J_2 efficiency of generation of new DNA breaks cannot be compensated by the process of their self-reparation.

We can see that the effect of radiation antagonism is, essentially, analogous to "hormesis" effect – if the latter satisfies the condition that an extremely weak ionizing radiation – "removes" consequences of ionizing effect of free radicals, the effect of radiation means that one extremely weak ionizing radiation "removes" consequences and effects of another, more intense, ionizing radiation.

Considered patterns and paradoxes are related to the case of "regular" water. But how does preliminary activation of water affect these anomalous radiobiological processes?

It is clear, that since with such activation the initial height of potential barrier V_0 is altered, the "hormesis" effect, as well as the effect of radiation antagonism will occur at slightly changed conditions.

In particular, since the condition for the "hormesis" phenomenon is determined by a special requirement to maximal intensity of weak ionizing field $J < WV_0/bk_b T$ (4.27), changing of variable V_0 in activated water corresponds to changing of this threshold.

Correspondingly, since the effect of radiation antagonism is fulfilled when $b_2 V_0/b_1 k_b T < J_1 \ll WV_0/b_1 k_b T$, changing of variable V_0 in activated water also changes the threshold and area of existence of this effect.

4.3 INFLUENCE OF CHARACTERISTICS OF HIGH-VISCOSITY FRICTION OF DNA MACROMOLECULES ON PECULIARITIES OF THEIR MOVEMENT AND PROCESS OF SELF-REPARATION IN REGULAR AND ACTIVATED LIQUID INNER CELLULAR MEDIA.

We can see from the conducted analysis that, in some cases, influence of one type of ionizing radiation (at its small intensity) may have radiation-generated protective effect with respect to action of another type of radiation with high intensity (for example, a relatively weak gamma-radiation may weaken destructive influence of an intense neutron radiation). There have been many experiments, in which this type of radiation-generated antagonism at combined influence of a neutron beam and a stream of gamma-radiation were confirmed. One of the possible reasons for such reciprocal radiation antagonism is in natural process of changing the characteristics of a biological object (particularly, inner cellular liquid in the space near a DNA molecule), when stimulation of non-enzymatic mechanism of self-reparation of double breaks of complimentary DNA chains takes place. The "hormesis" effect, for which a weak ionizing irradiation may play a radiation-based protective role with regard to influence of free radicals on DNA, has a similar explanation. These processes were investigated earlier. Moreover, since the process of DNA self-reparation occurs during movement of fragments of DNA in water-salt environment, the problem of inertness of such movement and influence of features of high-viscosity tension on it becomes important.

Solution of such problem will allow us to answer some logical questions.

How should both types of influence (the main one, causing DNA breaks and supplementary, which function consists of stimulating non-enzymatic mechanism of self-reparation) be reconciled? Is it necessary to have a constant "radio-protective" radiation or it can be applied in repetitive short bursts?

These are complicated questions and complete answers require understanding of many, still unknown, parameters. Meanwhile, some characteristic patterns may be established on the basis of simple mechanical models.

As an example, we shall consider features of movement of DNA fragments in real high-viscosity inter – molecular water-salt media under the influence of an initial short-duration impulse of radiation, applied as a burst. There may be several reasons for such initial impulse. The main reason is related to the fact that a DNA macro molecule is a complicated dynamic system with balanced links in a strained state (like prestressed concrete). One of the main elements of this rigid structure is the sugar-phosphate frame, holding together separate pairs of nucleotides. With a double break of elements of this frame, stressed condition is removed causing the initial burst.

In order to simplify the solution, we shall limit ourselves with considering the dependence of movement characteristics of biological macro molecules from media viscosity parameters. It is necessary to note, that we have considered one of the aspects of influence of media viscosity earlier, based on the assumption that viscosity of inner cellular media had a significant influence on dispersion characteristics of water molecules, and that, in its turn, affects the balance of

Coulomb forces and forces of Van Der Waals interaction of different pairs of nucleotides, located on both sides of a double DNA break. The problem considered here characterizes another side of influence of viscosity of inner cellular media on the process of radiation-generated protection of biological objects and allows to study the features of dynamics of DNA fragments' movement.

The force of high-viscosity tension of a molecule in real dense environment in case of a random degree-level rule of its dependence from velocity of movement $v = dx/dt$ may be written as

$$F_f = -k(dx/dt)^S \qquad (4.34)$$

where k – coefficient of tension, $S = m/n$, m and n – positive whole real numbers.

Equation for movement of a molecule has the form

$$M(dv/dx) = -k(v)^{(m/n)} \qquad (4.35)$$

This equation can be solved at initial conditions: $x = 0$, $dx/dt = v_0$ at $t = 0$.
Here M – molecule mass, v_0— initial velocity of a molecule.
Straight integration produces the result

$$V = \{((m - n)\ kt/nM) + v_0^{(n - m)/n}\}^{n/(n - m)} \qquad (4.36)$$

We shall consider separate special cases, which are defined by the character of dependence of the force of tension from velocity of movement of a molecule.

a)   Weak dependence of the force of tension from velocity. In this case $S < 1$, which is matched by the condition $m < n$.

In this case, from (4.36) we have

$$v(t) = v_0\ \{1 - (n - m)\ kt/v_0^{n/(n-m)}nM\}^{n/(n - m)} \qquad (4.37)$$

From (4.37) we find the expression for current coordinate of a molecule as a function of time elapsed after initiation of movement

$$x(t) = \int_0^t v(t)dt =$$
$$[nMv_0^{(2n - m)/(n - m)}/(2n - m)k]\ \{1-[1 - (n - m)kt/v_0^{n/(n-m)}/nM]^{(2n - m)/(n - m)}\} \qquad (4.38)$$

As we see, with increasing of time t, velocity $v(t)$ decreases and, at a moment of time

$$t_s = Mv_0^{n/(n-m)}(n/k(n - m)) \qquad (4.39)$$

a molecule stops at a point with coordinate

$$x_s = Mv_0^{(2n-m)/(n-m)} \, (n/(2n-m)k) \tag{4.40}$$

The same case also accommodates the mode of slowing down of particles with constant force of friction, independent of time, which is fulfilled by the condition $m = 0$. In that situation, the time of slowing down and final coordinate of a molecule are equal, respectively

$$t_s = Mv_0/k \tag{4.41}$$

$$x_s = Mv_0^2/2k \tag{4.42}$$

b)   Linear rule of high-viscosity tension, for which $S = 1$, $m = n$.

A direct transfer $m \to n$ in the general formula (4.36) for $v(t)$ is impossible due to appearing indeterminacy. First, we need to transform the formula for $v(t)$ so that we make a border transformation $m \to n$ in the item, independent of indeterminacy, while leaving the difference between coefficients $(n - m)$ in the first item, containing time t, in its explicit form. With such approach, we have

$$v(t) = v_0\{1 - (n - m)kt/nM\}^{n/(n-m)}$$

If we imagine the difference $(n - m)/n$ formally as a small value $(n - m)/n = \delta \to 0$, the final expression for velocity becomes

$$v(t) = v_0 \, \{1 - \delta \, kt/M\}^{1/\delta} \tag{4.43}$$

The limit for this expression at $\delta \to 0$ is an exponential function

$$v(t) = v_0 \, exp(-kt/M) \tag{4.44}$$

We see, that in this case, velocity of a molecule is continuously reduced, without becoming zero anywhere, which corresponds to a full stop only with $t \to \infty$.

From (4.44) we find the expression for current coordinate of a molecule

$$x(t) = v_0 \int_0^t e^{-kt/M} \, dt = v_0 M/k(1 - e^{-kt/M}) \tag{4.45}$$

We have noted with interest, that in this case, despite the fact that, according to (4.45), velocity never becomes zero, there exists an extreme point of movement (the point of stoppage)

$$x_s = v_0 M/k, \qquad (4.46)$$

depending on mass of a molecule M and initial velocity v_0.

c)   Strong dependence of the force of high-viscosity tension from velocity. This case provides for coefficient $S > 1$, $m > n$ in (4.24) for breaking force.

To analyze this situation, we need to use the revised formula (4.36) for velocity

$$V(t) = v_0/\{(m - n)k\ v_0^{(m-n)/n}\ t/nM + 1\}^{n/(m-n)}$$

We see that with $S > 1$, a molecule will not come to a full stop in a dense media over a finite period of time, while velocity will decrease asymptotically, becoming equal to zero only with $t \to \infty$.

From the expression for $v(t)$ we have

$$x(t) = \int_0^t v(t)dt = nMv_0^{-(m-2n)/n}/(2n - m)k\{1-[(m-n)kv_0^{(m-n)/n}t/nM+1]^{(m-2n)/(m-n)} \qquad (4.47)$$

Apparently, this case is the most interesting. It may be further divided into three possible variants depending on relation of parameters of non-linearity m and n in (4.34) for the force of high-viscosity tension.

1c) We shall first consider the case, when parameters of degree of non-linearity m and n, along with the initial expression $m > n$, satisfy the condition $2n > m$.

In this situation, indicator of the square parenthesis $(m - 2n)/(m - n)$ in (4.47) is negative. In result, from (4.47) we find

$$x = (nMv_0^{|2n-m|/m}/(2n - m)k)\{1 - 1/[|m - n|kv_0^{m-n/n}\ t/nM + 1]\}^{|2n-m|/|m-n|}$$

It can be seen that with an unlimited increase of the time breaking ($t \to \infty$), there is a finite point of stoppage of a molecule

$$x_s = nMv_0^{|2n-m|/m}/\ (2n - m)k \qquad (4.48)$$

2c) We shall now consider the situation, which emerges when $m > n$ and $m > 2n$.

In this case, the expression for $x(t)$ is characterized by the formula

$$x = nMv_0^{-|m-2n|/n} // (m - 2n)k \{[M - n]kV_0^{m-n/n} t/nM + 1]^{|m-2n|/m-n} - 1\}, \quad (4.49)$$

which means that relocation of a molecule becomes infinitely large, i.e. $x_s \to \infty$ at $t \to \infty$.

3c) We now conduct an analysis of the intermediate case $m > n$, $m = 2n$. This case cannot be investigated directly, using the finite transformation $m \to 2n$ for cases 1c) and 2c), because it would create an indeterminacy for x in the expressions obtained above

$$x \sim (1 - 1/A^{2n-m})/(2n - m)$$

It would be easier to find a solution, proceeding from the initial equation for movement (4.36), having assumed $m/n = 2$. Then, from (4.36) it would be easy to find velocity of movement

$$v(t) = v_0/(1 + kv_0t/M)$$

and expression for current location of a molecule

$$x(t) = \int_0^t v(t)dt = (M/k) \, ln(1 + kv_0t/M) \qquad (4.50)$$

As we can see, in this case there is also no final point of stoppage and the coordinate of a moving molecule increases infinitely with time (although an increase of the coordinate occurs very slowly due to logarithmical dependence of the coordinate from time).

This analysis shows that during movement of macro molecules in a media with gradual rule for the coefficient of high-viscosity tension $F = -kv^s$, it comes to a stop at an ultimate (from the starting position) distance only at $S < 2$. Stoppage occurs most quickly with a fractional ($S < 1$) indicator of dependence of the coefficient of high-viscosity tension from velocity. In the case, when a non-linear (in relation to velocity of movement) indicator of breaking is characterized by a stronger than quadratic gradual indicator $S > 2$, the character of movement changes principally – despite the presence of high-viscosity tension, movement becomes infinite and doesn't have a final stoppage point over a finite time period ($x \to \infty$ at $t \to \infty$).

The scenario for the process of self-reparation, considered above, shows that separate (in the result of double DNA breaks under the influence of hard gamma or neutron radiation) fragments of DNA may begin reciprocal rapprochement due to appearance of an inducing force, related to a defined change of dispersion features of an inner cellular liquid during an additional irradiation. Because of extreme weakness of that force, movements, contributing to restoration of DNA integrity happen slowly.

Clearly, from the point of view of radiation resistance of biological structures, the case of maximally rapid stoppage of macro molecules after a break (the closer it gets to stoppage, the easier it is to turn it back!) proves to be optimal.

Given the imperfect character of the force of high-viscosity tension (with the gradual rule with indicator $S \geq 2$) such movement leads to a very large and unmanageable separation of severed DNA fragments. Clearly, the mechanism of non-enzymatic self-elimination of breaks being considered is efficient only when spatial width of a break is not large and there are no elements of other DNA molecules between severed ends. Due to unavoidable entanglement of fragments of different DNA molecules on large distances, the process of self-elimination does not lead to elimination of a break, even after activation of the source of a defined change of dispersion features of an inner cellular media. Moreover, if high-viscosity tension doesn't stop a macro molecule fast enough, even a constantly applied additional irradiation cannot provide conditions for self-elimination of damage.

Therefore, accounting for specific features of influence of viscosity of inner cellular media on movement of biological macro molecules creates conditions for most efficient usage of the effect of radiation-based protection of DNA and biological systems in general under the influence of external ionizing radiation.

Nevertheless, it is important to note the following non-trivial circumstance. In the case of "regular" (i.e. non-activated) water, the expression for the force of high-viscosity tension corresponds to expression (4.34) with parameter of non-linearity $S = 1$. Deviation of S from one is possible in structured environments like water with macro clusters, for example, which are formed during its activation. Unfortunately, the problem describing the character of such high-viscosity tension in activated water is practically unresearched.

Therefore, accounting for peculiar features of influence of viscosity of inner cellular media on movement of biological macro molecules creates conditions for the most efficient usage of the effect of radiation-based protection for DNA and biological systems in general under the influence of external ionizing radiation.

Literature to Chapter 4

Vysotskii V.I., Pinchuk A.O., Kornilova A.A., Samoylenko I.I. // Radiation Physics and Chemistry, v. 65, n. 4-5, 2002, *p. 487-493.*

Zenin S.V., Tiaglov B.V. // J. of Physical Chemistry, v. 68 (1994), *p. 636.*

5. SPECIAL EFFECTS OF NON-IONIZED LOW FREQUENCY ELECTROMAGNETIC FIELDS ON BIOLOGICAL SYSTEMS IN ACTIVATED WATER

5.1 FEATURES OF SPATIAL STRUCTURE AND EFFICIENCY OF NON-IONIZING FIELDS ADOPTED FOR USE BY HUMANS IN CLOSE PROXIMITY TO SOURCES OF MAN-GENERATED RADIATION

5.1.1 Structural features of longitudinal and transverse non-ionizing fields in close proximity to a radiation source

It is well known, that presence of a broadband non-ionizing radiation of human origin is one of unavoidable results of civilization. Obviously, the issue of providing human safety in using non-ionizing fields should be regarded a top priority. There is a lot of evidence suggesting that exposure to such radiation disturbs normal functioning of human organism and initiates remote irreversible processes in many systems of the body. This problem becomes especially important when there is combined effect of ionizing and non-ionizing radiation as well as simultaneous effect of several other factors – for example, significant variations of temperature, presence of chemical factors altering the existing biochemical balance in the organism, the stress factor and so forth.

It needs to be noted that usually characteristics of negative effects of non-ionizing electromagnetic waves are attributed to their power and frequency. However, as we are going to show here, wave power is not the characteristic, which, along with frequency, can unambiguously determine efficiency of influence on a biological object. Apparently, wave power (and wave intensity, directly linked to it) of an electromagnetic wave determines properties of an electromagnetic field in the remote (wave) zone. Moreover, as a simple analysis shows, people most often find themselves in the near "non-wave" zone of a consumer or commercial source of radiation adopted for human use. There, such a common parameter of any source of radiation as its power cannot definitely describe properties of an electromagnetic field. That may seriously challenge traditional ways of assessing radiation safety.

Let's consider characteristics and features of impact in the near (non-wave) zone of those longitudinal and transverse non-ionizing fields and waves, which are equivalent to low power radiation in the remote (wave) zone.

For clarity, we shall consider the problem based on the following model.

Suppose there is a source of electromagnetic radiation with frequency w and a relatively low power P_0. It is commonly believed that knowing the power of radiation and distance from the source to an object of radiation would be enough to characterize its effect. We shall demonstrate that by doing it conventionally we may get a very big mistake and seriously underestimate the actual effect of radiation from a source being in fact much more powerful.

Using the standard expression for calculating the electric Hertz vector we calculate the field

$$\vec{\Pi}(\vec{R},t) = (1/\varepsilon) \int_V \vec{p}(\vec{R}_l,t) \frac{e^{ikr}}{r} dV_l \qquad (5.1)$$

Here ε - dielectric permittivity of medium with frequency corresponding to radiation, $\vec{r} = \vec{R} \pm \vec{R}_1$ – distance from a specific spot in the volume of the source $\vec{R}_1$ to a point of observation $\vec{R}$, $\vec{p}(\vec{R}_1, t) = \vec{j}/i\omega$ – oscillating dipole moment related to current with density of $\vec{j} = \partial \vec{p}/dt$ being source of the field.

Integration is performed on the full volume used by the current.

Let's consider the most important case of non-ionizing radiation, when the field source is located within the area much smaller than wavelength of radiation. In very many actual systems it is accomplished by a big margin (for example, mobile phones have frequency of radiation 900-1900 *MHz*, which corresponds to a wave length of about 16-30 *sm*, which is much higher than the size of transmitting antenna with a length of no more than 4-5 *sm*). In this case we may use the following expansion

$$r = | \vec{R} - \vec{R}_1 | \approx | \vec{R} | - \vec{e}_{\vec{R}} \vec{R}_1 \qquad (5.2)$$

where $\vec{e}_{\vec{R}}$ - singular vector pointed towards the radius-vector of the point of observation $\vec{R}$. Absolute value R is equal to distance from point of observation to the center of the area, where sources of radiation are located. Expanding the exponent in the denominator of the expression for the Hertz vector, we find

$$e^{ikr}/r = \frac{e^{ikR}}{R}[1 + (1/R - ik)\vec{e}_{\vec{R}}\vec{R}_1 - (1/2)(k^2 + 2ik/r - 2/R^2)(\vec{e}_{\vec{R}}\vec{R}_1)^2] \qquad (5.3)$$

Clearly, there is a consistent account of contributions made by different areas of dislocation of sources on outcome radiation. The first element of this expansion matches the ultimately small relation of dimensions of the source to wavelength.

As a result, using only the first item from the last expression, we have

$$\vec{\Pi}(\vec{R},t) = (e^{ikR}/\varepsilon R) \int_V \vec{p}(\vec{R}_1,t)dV_1 = \frac{e^{ikr}}{\varepsilon R}\vec{p}_0 \qquad (5.4)$$

Here p_0 – full dipole moment of the irradiating system.

Using the obtained expression for the Hertz vector it is possible to find vectors of electric and magnetic fields created by this source

$$\vec{E} = grad\,div\vec{\Pi} + k^2\varepsilon\mu\vec{\Pi};$$
$$\vec{H} = -ik\varepsilon\,rot\vec{\Pi}$$

(5.5)

Orienting the vector of the dipole moment p_0 along the axis oz of the spherical system of coordinates (see *Fig. 1*), from (5.5) we can obtain expressions for items of the electromagnetic field, different from zero

$$E_R = 2(p_0/\varepsilon)\frac{e^{ikr}}{R^3}(1-ikR)\cos\vartheta,$$

$$E_\vartheta = (p_0/\varepsilon)\frac{e^{ikr}}{R^3}(1-ikR-k^2R^2\varepsilon\mu)\sin\vartheta,$$

$$H_\varphi = -ikp_0\frac{e^{ikr}}{R^2}(1-ikR)\sin\vartheta,$$

$$E_\varphi = H_R = H_\vartheta = 0$$

(5.6)

In this expression the component representing strength of the electric field E_R is directed along the radius vector R and corresponds to the longitudinal component. The other two items of the electric and magnetic fields, different from zero, are perpendicular relative to R.

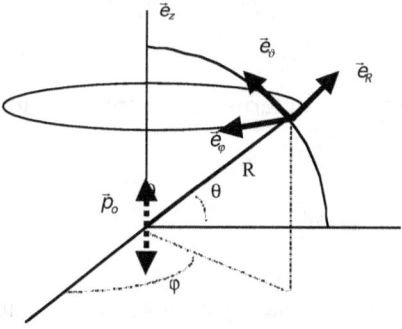

Fig. 5.1   Mutual location of irradiating dipole moment $\vec{p}_0$ and other separate components of the field

In the near wave zone (at $R \ll \lambda$ and $kR \ll 1$) field components (6) assume a simplified form

$$E_R = 2(p_0/\varepsilon)\frac{1}{R^3}\cos\vartheta,$$

$$E_\vartheta = (p_0/\varepsilon)\frac{1}{R^3}\sin\vartheta,$$

$$H_\varphi = -ikp_0\frac{1}{R^2}\sin\vartheta, \qquad\qquad (5.7)$$

$$E_\varphi = H_R = H_\vartheta = 0$$

We can see that in the near zone the amplitude of the longitudinal field component E_R is twice bigger than amplitude of the perpendicular component $E\varphi$. We can also see that in this area the strength of the magnetic field is by $1/kR \gg 1$ times smaller than both components of the electric field E_R and $E\varphi$, different from zero.

5.1.2 Efficiency analysis of the effect of ionizing fields in the near zone of sources of radiation adapted for human use

We shall calculate intensity of an electromagnetic field in the near zone from a source of radiation.

We shall use the general expression for density of energy stream (intensity) of an electromagnetic field

$$J = \frac{c}{8\pi}\,|\,[\vec{E}\times\vec{H}]\,| = \frac{c}{8\pi}\{|\,E_R H_\varphi\,| + |\,E_\vartheta H_\varphi\,|\} \qquad\qquad (5.8)$$

In this expression the first item determines energy stream of a field

$$\vec{Y}_1 = Y_\vartheta\vec{e}_\vartheta = \vec{e}_\vartheta\frac{c}{8\pi}\{E_R H_\varphi\} \qquad\qquad (5.9)$$

in the direction of polar angle Θ (in the polar system of coordinates it represents transformation of energy of a field from one pole to another).

The second item

$$\vec{Y}_2 = Y_R\vec{e}_R = \vec{e}_R\frac{c}{8\pi}\{E_\vartheta H_\varphi\} \qquad\qquad (5.10)$$

characterizes transformation of energy of a field along the radius (i.e. in the direction away from the source). We shall find that part of power, which corresponds to the second item.

In the case of a non-magnetic environment with $\varepsilon = 1$ we have

$$J_{R \ll \lambda} = (p_0^2 c / 4R^5 \lambda) \sin^2 \theta \qquad (5.11)$$

Total power of an electromagnetic field corresponding to that component in the near zone can be found by integration of $J_R \ll \lambda$ on the entire spherical surface with radius $R \ll \lambda$, fully inclusive of the system of currents

$$P_{R \ll \lambda} = \int J_{R \ll \lambda} dS = (4\pi p_0^2 c / 3R^3 \lambda) \qquad (5.12)$$

We can see that this power decreases with growing distance and $P_{R \ll \lambda} \sim 1/R^3$.

We should note that although this field is not directly connected with processes of radiation (for that total power should not decrease with growing distance from a source), but it actually exists, affects the movement of electrons and ions and may influence objects located in the area of field's existence.

Correspondingly, in the remote (wave) zone (at $R \gg \lambda$ and $kR \gg 1$) we have a principally different result

$$
\begin{aligned}
E_R &= -2ik(p_0/\varepsilon)\frac{e^{ikr}}{R^2}\cos\vartheta, \\
E_\vartheta &= -k^2 \mu p_0 \frac{e^{ikr}}{R}\sin\vartheta, \\
H_\varphi &= -k^2 p_0 \frac{\sqrt{\varepsilon\mu}\, e^{ikr}}{R}\sin\vartheta, \\
E_\varphi &= H_R = H_\vartheta = 0
\end{aligned}
\qquad (5.13)
$$

Using the same general expression for density of an energy stream (intensity) of an electromagnetic field

$$J = \frac{c}{8\pi} \mid [\vec{E} \times \vec{H}] \mid = \frac{c}{8\pi} \mid E_\vartheta H_\varphi \mid$$

we find (in a non-magnetic environment with $\mu = 1$ and $\varepsilon = 1$) an expression for intensity

$$J_{R \gg \lambda} = (2\pi^3 c\, p_0^2 / R^2 \lambda^4) \sin^2 \theta \qquad (5.14)$$

Full power of radiation in this case can also be found by integration by surface and is equal

$$P_{R\gg\lambda} \equiv P_0 = \int J_{R\gg\lambda} dS = (16\pi^4 c\, p_0^2/3\lambda^4) \qquad (5.15)$$

We can see that power doesn't depend on distance and therefore such field matches an actual electromagnetic field of radiation with power P_0. From (5.15) we can determine relation of power P_0 with value of dipole moment

$$p_0 = (3\lambda^4 P_0/16\pi^4 c)^{1/2} \qquad (5.16)$$

and the finalized expression for power in the near zone

$$P_{R\ll\lambda} = 2(\lambda/2\pi R)^3 \, P_0 \qquad (5.17)$$

On the basis of (5.12) and (5.17) it is possible to compare powers of an electromagnetic field in different zones

$$K = P_{R\ll\lambda}/P_{R\gg\lambda} \approx 2(\lambda/2\pi R)^3 \qquad (5.18)$$

It follows from the finalized relation (5.18) that in the near ("non-wave") zone at distance $R < \lambda/2\pi$ from a source of radiation effective power $P_{R\ll\lambda}$ of an electromagnetic field exceeds respective power of radiation field in the remote ("wave") zone $P_{R\gg\lambda} \equiv P_0$ (see *Fig. 5.2*).

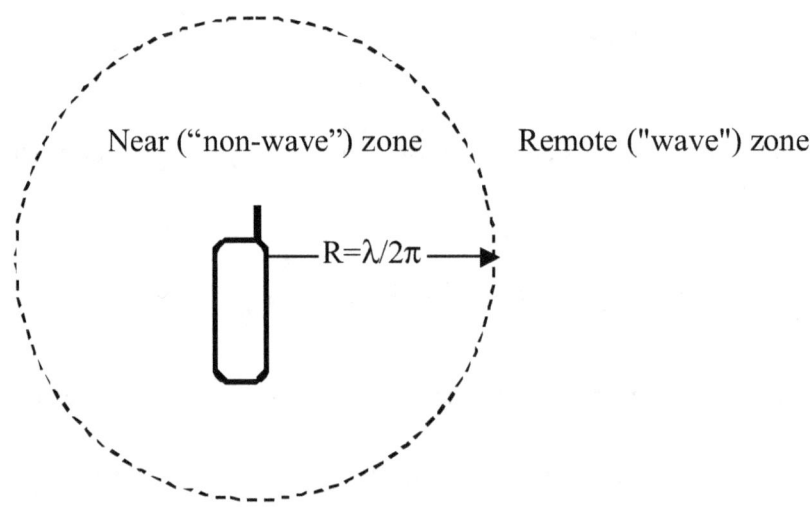

Fig. 5.2 Near ("non-wave") and remote ("wave") zones of radiation source with wavelength λ

Let's consider possible consequences of anomalous increase of effective intensity of an electromagnetic field, which occur in the near zone of a radiation source.

As we well know from electrodynamics, the effect of non-ionizing fields may manifest itself, first of all, through their influence on currents and charges. These charges are particularly numerous in biological systems, where they are contained in inner cellular water-salt medium, being in essence an electrolyte or a dense partially ionized electronic/ionic plasma, which is identical to the latter. Due to that circumstance, it doesn't really matter whether there is an electromagnetic field in the waveform, which has "broken away" from its source, or it is an electromagnetic field inseparable from source. Both types of a field affect currents the same way. For that reason, the decisive factor for estimating efficiency of effect of a field on a biological object is its intensity. It allows to compare efficiency of an effect of the same source by comparing intensities of fields (5.7) and (5.13) in the near and far zones as well as effective powers (5.18) in these areas of space.

Here are some numeric assessments related to sources of non-ionizing radiation maximally adapted for human use.

For example, in a field created by a powerful radio station with broadcast frequency 100 kHz (wavelength $\lambda = 3$ km) and power $P_0 = 100$ kW effective power in the near zone at the distance of 100 m from transmitting aerial increases by 250 times in comparison with its specification power corresponding to power of radiation in the far zone, being equal to $P_{R<<\lambda} = 25$ MW.

Here is another example. A mobile phone has a small specification power of radiation $P_0 = 0.5$ W and works in the frequency range of 900 MHz with corresponding wavelength of $\lambda = 30$ sm. According to its specifications such mobile phone is considered safe (one of the accepted criteria defines the safety threshold by the power density of its effect on the brain of about 1 W/kg of biological tissue of the brain).

If a user holds a phone at the distance of $R = 3$–5 cm from sensitive organs of the human brain its effective (actual) power increases by 432–2000 times and reaches $P_{R<<\lambda} = 2(\lambda/2\pi R)^3 P_0$ = 216 – 1000 W.

In that case power density of an electromagnetic field exceeds the value of 200–1000 W/kg, which is over the normative safety threshold by 200–1000 times!

Similar estimates can be made with regard to such a popular source of non-ionizing SHF radiation as monitor of a personal computer.

We can offer some additional information related to the effect of the magnetic component of an electromagnetic field on biological objects.

Using expressions (5.7) and (5.16) it is easy to obtain the expression for relation of effective strength of a magnetic field in the near zone of radiation source from its "specification" power P_0 of that source

$$H_\varphi = (\lambda/2\pi R^2)(3P_0/c)^{1/2} \qquad (5.19)$$

For the example of a powerful long-wave radio station with $P_0 = 100$ *kW*, strength of an alternating magnetic field at the distance of $R = 100$ *m*, $P_{R \ll \lambda} = 25$ *MW* corresponding to its effective power is equal to $H_\varphi = 0.02$ Oersted.

Similarly, for a mobile phone with power $P_0 = 0.5$ *W* and working frequency 900 *MHz*, strength of a magnetic field at the distance $R = 3$–5 *sm* from transmitting antenna (i.e. in the area of the brain) is equal to $H_\varphi = (0.03$–$0.1)$ Oersted.

How should we perceive that value? Is it high or low?

The easiest way is to compare it with the standards used by reputable organizations. For example, according to the radiation safety standards used in Sweden, the maximum allowed strength of a low-frequency magnetic field is equal $H_{max} = 0.002$ Oersted. For a high-frequency magnetic field requirements are even stricter.

We can see that strength of a magnetic field created in the "near" zone around the antenna of a mobile phone exceeds this safety standard by 15–50 times!

We should note that in the "near" zone of a radiation source (at distance $R \ll \lambda$) standard radiometric equipment is not effective. Those devices are designed for work in the "far" wave zone with $R \gg \lambda$. As a result, a very significant amplification of the effect of radiation sources in the near zone is hard to register, which makes situation worse.

Interestingly, if an assessment of strength of an alternating magnetic field of a mobile phone transmitter is performed at the same distance R from the antenna, using the regular expression for relation of power of radiation from strength of a magnetic field, which would be true for the remote (far) wave zone

$$H = (4P_0/c)^{1/2}/R \qquad (5.20)$$

with the same values $P_0 = 0.5$ *W* and $R = 3$–5 *cm*, we have $H = (0.009$-$0.015)$ Oersted, which is by 3.5–6 times less than it really is.

The simplest way to lower such effect in the near zone is to place the antenna away from the body of a mobile phone and, possibly, increase working frequency of radiation (reduce wave length), but in that case other problems may arise related to effects of a higher frequency radiation, which may become close to natural resonance of absorption of water and biological molecules. Other ways of optimization are possible related to controlled reduction of sensitivity of biological tissue to ionizing influence and development of methods of changing structure of an electromagnetic field in the "near" zone.

5.2 PHYSICAL MECHANISMS OF INFLUENCE OF NON-IONIZING FIELDS ON RESISTANCE OF BIOLOGICAL OBJECTS TO RADIATION

5.2.1 Problem of influence of non-ionizing fields on biological objects

It is well known that presence of ionizing irradiation has a specific effect on stability of any biological object. The analysis conducted earlier has demonstrated that depending on the character of that radiation (its energy, duration, intensity and other factors) and features of a biological objects itself ionizing irradiation may cause both irreversible destruction (degradation) of information macro molecules of DNA, which are the vital element of any living organism, and stimulation of their stability (including stability due to non-fermentation mechanisms at for example, compensatory destructive effect of free radicals of non-radiation origin). It is important to note that such features are the direct result of the very specifics of the effect of ionizing radiation – by definition it's ionizing, causing the ionic content of atoms and molecules, formation of free radicals.

A logical question arises – can non-ionizing radiation or influence of other non-ionizing fields affect stability of biological macromolecules and if so, how? This question is important because non-ionizing irradiation is the factor of external influence, which is always present in the environment and affects any biological object. The presence of irradiation with the broadest specter of radio frequencies is an indispensable man-generated result of civilization. The influence of such radiation is the strongest in radio locator systems and powerful means of communication.

The fact that such influence exists is beyond doubt. It has been proven in numerous experiments. Normally, the mechanism of influence of non-ionizing fields is related to thermal effect. Incompleteness and often even fallacy of such approach is evident. For example, strong influence of stationary (or slowly changing) magnetic fields on biological objects is well known. However, thermal influence of a stationary magnetic field on a non-magnetic system is always so small that it can be ignored. Small heating-up of any object containing water will also occur under the effect of fields with radio frequencies, different from frequencies of resonance absorption (main lines of absorption in water in the centimeter and millimeter frequency ranges corresponds to frequencies of 22 *GHz* and 183 *GHz*). Based on the specific heat value of $C = 4.2$ *J/g* grad for water and taking into account that physiological consequences of over-heating may manifest themselves beginning from changing the temperature of internal organs by $\Delta T \geq 1$ grad (smaller temperature changes do not significantly affect living organisms) it is easy to verify that influence of the thermal mechanism of non-ionizing irradiation should be registered during absorption of energy exceeding 240 *J* per 1 *kg* of a living system (at the condition of its complete thermal isolation). Allowing for an unavoidable loss of heat due to natural thermal regulation of living objects over the control period of 1 minute the threshold power density of energy absorption of non-ionizing irradiation should be more than $P_0 = 4\text{-}5$ *W* per 1 kilogram of a living system. Such a big threshold value is significantly higher than power density absorbed from typical sources of non-ionizing irradiation.

At the same time, there is a lot of reliable evidence suggesting a very significant negative effect of a relatively weak non-ionizing radiation with $P \ll P_0$ on living systems. At the other hand, there are also reliable facts of a positive effect of such irradiation causing, for example, a significant weakening of the destructive effect accompanying ionizing radiation, which presents a kind of radiation antagonism. Obviously, such facts cannot be explained by thermal mechanism.

Below, there are some possible mechanisms of non-thermal influence of non-ionizing radiation on stability of any biological object, considered for the first time.

5.2.2   Mechanisms of dispersion influence of non-ionizing electromagnetic fields on DNA stability

We have shown in our previous works that the problem of DNA stability with respect to double breaks – the most dangerous mechanism of degradation – can be solved by an analysis of features of power influence of severed DNA fragments in the area of a double break. Here, the main factor affecting efficiency of the process of self-reparation of a double break is presence and height of a potential barrier appearing in the area of a break. Parameters of a break depend on the Coulomb interaction between charges distributed along the surface of those nucleotides, which are located on both sides of a double DNA break

$$V^Q = \sum_{i=1}^{N_\alpha} \sum_{j=1}^{N_\beta} V^Q_{i,j} \qquad (5.21)$$

and on the long-acting dispersion interaction of the Van Der Waals type

$$V^{vdw} = \sum_{i=1}^{2} \sum_{j=1}^{2} V^{wdv}_{i,j} \qquad (5.22)$$

located between the nucleotide pairs on opposite sides of a DNA spiral break.

Here $V^{vdw}_{i,j}$ - energy of dispersion interaction between separate nucleotides i and j located on opposite ends of a double DNA spiral break, N_α, N_β - number of atoms of a respective complementary pair of nucleotides GC (guanine-cytosine) or AT (adenine-thimine) ($N_\alpha = 29$ for GC, $N_\beta = 27$ for AT); $U^Q_{i,j}$ – energy of pair Coulomb interactions of charges Q_i and Q_j distributed along the surface of a nucleotide pair GC or AT under provision of real geometry of distribution of fractional charges, while the variable

$$V^{vdw}_{jk}(R) = -\frac{27\eta}{16\pi^3} \frac{V_k V_j}{R^6} \int_0^\infty \frac{\left(\varepsilon_j(i\omega) - \varepsilon_3(i\omega)\right)\left(\varepsilon_k(i\omega) - \varepsilon_3(i\omega)\right)}{\left(\varepsilon_j(i\omega) + 2\varepsilon_3(i\omega)\right)\left(\varepsilon_k(i\omega) + 2\varepsilon_3(i\omega)\right)} d\omega$$

determines energy of interaction of the Van Der Waals type and depends on respective dielectric permittivity of each of the interacting objects (nucleotides) $\varepsilon_j (i\omega)$, $\varepsilon_k (i\omega)$ as well as on dielectric permittivity of the inner cellular medium $\varepsilon_3 (i\omega)$, where DNA is located.

Normally, energy of Coulomb interaction (5.21) is found based on the Debye-Gukkel theory, which takes into account processes of self-regulated screening of a field of nucleotide charges by ions contained in the inner cellular liquid. To find such a screened field, we need to solve the linearized equation of Puasson-Bolzman

$$\nabla^2 \tilde{\Psi} = g^2 \tilde{\Psi} \qquad (5.23)$$

which was thoroughly investigated earlier.

Here, $g^2 = 8\pi n e^2 / \varepsilon_3(0) k_b T$, g – the Debye constant, equal to radius of screening of charges in plasma, $n = \frac{1}{2} \sum_z n_z^0 z^2$ – ionic strength of a solution.

For a neutral solution the full charge of the system is equal to zero $\sum_z n_z^0 z = 0$.

This results leads to a situation, when a potential in the area between these charges shaped like spheres with radiuses a_1 and a_2 will be expressed by the sum of screened potentials of separate spheres $\tilde{\Psi}_1 + \tilde{\Psi}_2$ each one of those having the form

$$\tilde{\Psi}(r) = \frac{q}{r} e^{-gr} \qquad (5.24)$$

Screening causes a sharp weakening of the Coulomb field beyond the limits of the radius of screening g^{-1}.

On *Fig. 5.3* there is the earlier calculated relation of energy of an electrostatic (curve 1) and electrodynamic (curve 2) interaction between end pairs of nucleotides (guanine-cytosine) – (guanine-cytosine) located on both sides of a double DNA break according to distance between them r, i.e. the width of a double break. The balance (curve 3) of the two main types of energy of interaction causes a situation, when a potential barrier emerges in the area of a double break.

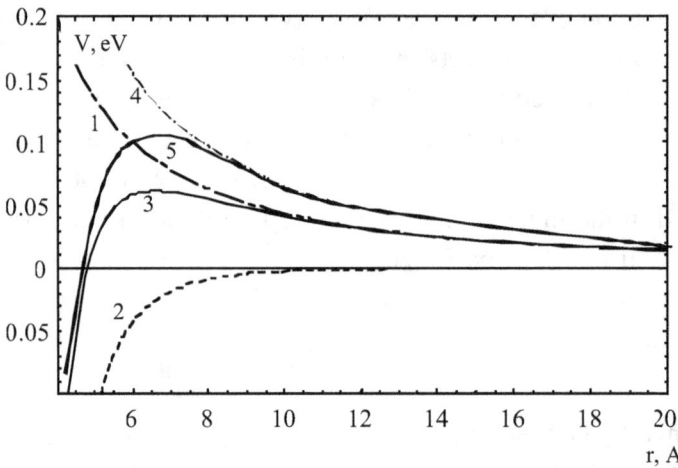

Fig. 5.3   Energy of electrostatic interaction (curves 1 and 4), energy of electrodynamic interaction (curve 2) and total energy of interaction (curves 3 and 5) between end pairs (guanine-cytosine) – (guanine-cytosine) depending to the width of a double break r. Curve 4 represents a more powerful Coulomb interaction and, eventually, a higher anti-reparation barrier 5.

In previous works we have conducted an analysis of special features of that barrier, investigated its influence on result movement of DNA fragments and considered results of an influence by an external ionizing irradiation on changing that barrier (mainly due to changing of the ionic content of the inner cellular medium).

Let's consider how ionizing irradiation influences the process of possible DNA self-reparation.

In order to do that we need to recall that the equation of Puasson-Bolzman (5.23) in the method of Del Re and Hukkel, as well as existence of screening of charges in a plasma-like environment, which follows from that equation, was obtained under a normal condition that the field of all charges is purely potential and that charges in plasma do not create a magnetic field. That condition holds only when all participating in interaction charges are static. Implicitly, it is also believed that there is no external magnetic field in an environment at all.

Obviously, there is some idealization here. Having an external slowly changing non-ionizing electromagnetic field in place, it will affect charges in liquid. Low-inertion quasi-free hydrated electrons will be affected most efficiently.

First of all, we need to determine whether a non-ionizing electromagnetic field can penetrate into a volume of a live organism consisting, mainly, of an inner cellular water-salt physiological solution. That solution contains a large number of heavy ions and hydrated electrons and, in effect, is a low-ionized plasma.

Dielectric permittivity of plasma is equal to

$$\varepsilon(\omega) = 1 - \omega_i^2 / \omega^2 - \omega_e^2 / \omega^2 \qquad (5.25)$$

Concentration of heavy ions and electrons in that environment assumes values close to $n_i \approx n_i \approx 10^{13}$ sm^{-3} (at a neutral value of pH 7). In such plasma there may exist plasma vibrations of the electronic and ionic subsystems. A given concentration of electrons and ions (in case they were completely free) is matched by plasma frequencies

$$\omega_e = \sqrt{4\pi n_e e^2 / m} \approx 1,5.10^{11}\, Hz$$

$$\omega_i = \sqrt{4\pi n_e e^2 / M_i} \approx 10^9\, Hz \qquad (5.26)$$

We know from electrodynamics that an electromagnetic wave can penetrate into plasma only when its frequency ω is higher than the highest one of the plasma frequencies (6), in this case electronic frequency ω_e).

If frequencies of an electromagnetic field are less than plasma frequencies, a field only partially penetrates its volume to the depth of its skin-layer, equal (independently of frequency of an electromagnetic wave) to the wavelength of plasma vibrations

$$\lambda \approx c / \omega_{e,i} \approx 0,2 - 30\, sm$$

It follows from this data that low-frequency non-ionizing fields can penetrate the volume of biomass to a considerable depth and affect the movement of charges. We shall examine that movement more elaborately.

Movement of each electron in an alternating field

$$\overset{\perp}{E} = \overset{\perp}{E}_0 e^{i(\overset{\perp}{k}\overset{\perp}{r}-\omega t)}$$

causes a periodic acceleration of electrons by a field and their braking by colliding with atoms and molecules of a liquid. An electron regulated by a field goes through f_{ea} collisions in 1 sec.

Frequency of collisions may be determined from an obvious relationship

$$f_{ea} = \sigma_a n_a <v_T> = \sigma_a n_a (3k_B T/m)^{1/2} \qquad (5.27)$$

Here, $<V_T>$ – average quadratic thermal speed of electrons, σ_a and n_a – cross section and concentration of neutral molecules in a liquid.

At the temperature $T = 319\, K$ (it is equivalent to $T = 37°C$), average frequency (5.27) of collisions of an electron with molecules of water equals $f_{ea} = 10^{14}\, s^{-1}$, which means that between two consecutive collisions an electron covers the distance of about 10 A.

Since the period of collisions $1/f_{ea} = 10^{-14}\, s$ is by many orders shorter than the period of non-ionizing electromagnetic wave ($T = 10^{-8}–10^{-11}\, s$), there will be an adiabatic process – the average drifting speed of an electron will change proportionately to instantaneous strength of an electromagnetic wave's field.

During each collision it loses impulse mv, where v – directed drifting speed of an electron in the direction of the field of an electromagnetic field. During time Δt an electron loses impulse

$$\Delta P = mv\, f_{ea}\Delta t$$

From this expression we can determine the average braking force

$$F = \Delta P/\Delta t = mv\, f_{ea} \qquad (5.28)$$

If we equate this braking force to the force, with which an electromagnetic wave $\overset{\perp}{E} = \overset{\perp}{E}_0 e^{i(\overset{\perp}{k}\overset{\perp}{r}-\omega t)}$ affects the same electron $F_E = eE$, we will be able to find an expression for average drifting speed of an electron in a wave's field

$$v = eE/\, mf_{ea} = eE/[\sigma_a n_a (3mk_B T)^{1/2}] \qquad (5.29)$$

Strength of a wave field is related to its intensity by the expression

$$J = E^2 c/4\pi\varepsilon(\omega)$$

Now, we find another expression describing relation of wave intensity to average drifting speed of electrons in the inner cellular liquid medium

$$v = e[4\pi J\varepsilon(\omega)/3mck_B T]^{1/2}/\sigma_a n_a \qquad (5.30)$$

When we have a field with frequency 100 MHz and intensity 1 W/sm^2 provided that dielectric permittivity of water on this frequency $\varepsilon(\omega) \approx 10$ we find that shear speed is equal $v = 10^3$ cm/s.

What would be different if charges in the inner cellular liquid moved with a periodically changing (synchronized with external alternating current) drifting speed v?

Potential of each charge will be different from its expression in (5.24), which describes static electrons, as it will assume a principally different form (Ignatiev, 1982)

$$\tilde{\Psi}(r) \approx \frac{q}{r}[e^{-gr} + (v/c)^2(1+\cos^2\theta)], \cos\theta = (\overset{\perp}{v}\overset{\perp}{r})/|\overset{\perp}{v}\overset{\perp}{r}| \qquad (5.31)$$

Here Θ - angle between the direction r from a charge to a point of observation and the vector of speed v of directed drift of electrons.

We can see that in this case there is no screening of the Coulomb field and it fades very slowly with distance. If we take into account that electrons periodically change direction and value of speed, the finalized expression for potential would have the form

$$\tilde{\Psi}(r) \approx \frac{q}{r}[e^{-gr} + <(v/c)^2 > (1+ <\cos^2\theta >)] \qquad (5.32)$$

As we can see, with increasing drifting speed of an electron the law of interaction between charges located on the surface of biomolecules becomes principally different – except the screened interaction $\tilde{\Psi}(r) = \frac{q}{r}e^{-gr}$, which decreases very rapidly with increasing distance due to the phenomenon of screening of charges, there will also be an additional kind of interaction, which would decrease very slowly with larger distance. Particularly, from (5.32) follows that beginning from the distance

$$r_{cr} \approx g^{-1}\ln(c^2 / <v>^2) >)$$

that additional interaction becomes the primary one. Weighted contribution of this additional item (according to expression (5.30)) is proportional to intensity of an electromagnetic wave, so this effect may be manifested in strong non-ionizing fields.

Another possibility of influencing interaction of biological macromolecules in an inner cellular liquid medium owes to the effect of an external magnetic field with induction B. As we noted earlier, the standard expression for screening in the Coulomb interaction is valid only under the condition that all interactions in a given system are of potential nature (i.e. energy may be expressed by an integral of work by several forces). A Coulomb electrostatic field satisfies that condition since in this case

$$V(r) = -\int_{r_0}^{r} F(r)dr$$

A magnetic field is not potential and its presence principally changes the character of interaction between charges. Overlapping of a static magnetic field on a biological object causes rearrangement and modulation of thermal movement of ions in an inner cellular water-salt environment. Transverse motion (with respect to direction of a magnetic field) of electrons becomes quantum-like and restricted. These effects significantly alter the features of screening and interaction of charges of DNA nucleotides. If in a volume of a live organism there is an external magnetic field with induction B affecting electrons in the inner cellular liquid, the expression for potential of any charge is different from its form (4) and is equal (Ignatiev, 1982)

$$\tilde{\Psi}(r) \approx \frac{q}{r}e^{-gr} + \beta g^2 \frac{q}{r^3}(1 - 3\cos^2\vartheta) \qquad (5.33)$$

Here

$$\beta = (\frac{\eta}{2mcg})^2 (\frac{1}{\varsigma} - \frac{1}{sh\varsigma ch\varsigma}), \varsigma = \eta \omega_B / 2k_B T, \omega_B = eB / mc, cos\theta = (\overset{r}{B}\overset{v}{r}) / |\overset{r}{B}\overset{v}{r}|$$

We can see that in this case there is also a very significant difference of the potential from the case when there is no magnetic field present.

Expression (5.33) is valid for any value of induction of a magnetic field, provided that the non-dimensional parameter β may become relatively big given a strong magnetic field (including $\beta \geq 1$).

It follows from these calculations that presence of a non-ionizing alternating electromagnetic field or a static magnetic field can principally change all relationships characterizing Coulomb interaction of nucleotide charges in the area of a double DNA break.

When an external non-ionizing electromagnetic field is present the regular law of Coulomb screening (4) representing a weaker reciprocal Coulomb repulsion of identical pairs of nucleotides on both ends of a double DNA break (curve 1 on *Fig. 5.3*) is replaced by expression (5.2). That expression describes a more powerful interaction (curve 4) causing intensification of the Coulomb repulsion at a constant dispersion attraction (curve 2), which eventually leads to a higher anti-reparation barrier (curve 5). Therefore, conditions for existence of the phenomena of radiation antagonism and hormesis examined earlier would be different. It may also create probability of inducing the effect similar to radiation synergy (i.e. mutual amplification of impact of degradation of the ionizing factor of DNA and non-ionizing field), opposite to radiation antagonism and hormesis (their reciprocal compensation).

In the case of a static magnetic field, replacing of screened Coulomb interaction (4) by expression (13) may cause both increasing of the Coulomb repulsion (in the interval of angles, where $(1 - 3cos^2\Theta) \geq 0$) and its partial or total suppression at $(1 - 3cos^2\Theta) \geq 0$) with a constant dispersion attraction. The nature of this effect is quite intricately related to the size of a magnetic field and distance between interaction ions. The average value of an additional (i.e. directly related to a magnetic field) component

$$< \Delta\tilde{\Psi}(r) >= \beta g^2 q < \frac{1}{r^3}(1 - 3cos^2\theta) >$$

of potential $\tilde{\Psi}(r)$ depends on mutual position of charges on the surface of nucleotides. Moreover, at a short distance $r << g$, given that $\beta \geq 1$, which is true for the case of a powerful magnetic field, this additional component may considerably exceed standard screen potential $\tilde{\Psi}(r) = \frac{q}{r}e^{-gr}$ and be positive as well as negative. These conditions may either suppress or amplify the effect of Coulomb interaction of nucleotides. In result, overlapping of a magnetic field on a biological object may cause both hormesis-type effects and radiation antagonism (i.e. lead to weakening of the impact of free radicals and an external ionizing radiation on DNA) as well as radiation synergy effects (i.e. stronger impact). Separate analysis is required for each case.

5.3   GENERAL PRINCIPLES OF INTERACTION OF NON-IONIZING MICROWAVE RADIATION WITH BIOLOGICAL SYSTEMS

Understanding of processes occurring in structures of the water phase of cells and tissues of live organisms under impact of microwave radiation is impossible to achieve without comprehension of all features of influence of an electromagnetic radiation on live systems. The main mechanism of an intense radiation is thermal effect. Most known hypothesis about mechanisms of influence of a weak microwave radiation on an organism in general is based on two main starting points:

1.   The primary target for a non-ionizing microwave radiation is water molecules related to protein systems of the skin collagen;

2.   Water molecules, which generally absorb microwaves, have an effect on receptor proteins of a cell's membranes with consecutive triggering of biochemical processes in a live organism.

Due to a multi-level structure of live systems, investigation of structural changes of the water phase of cells and tissues occurring under influence of an electromagnetic radiation is a very complicated task. Its full solution is still in the future. We can point out some main points characterizing the nature of such interaction:

1.   Microwave radiation energy absorbed by molecules before degradation into heat can transform physical and chemical features of complex biological molecules.

2.   All biological objects contain water, which, in general, absorbs energy of a microwave electromagnetic radiation. The main mechanism here is volume absorption.

3.   Using multi-level radiation usually increases efficiency.

4.   Microwave radiation has comparatively low frequency and therefore can alter only rotational conditions of molecules.

5.   At normal temperatures, water molecules form a grid structure of hydrogen links constantly changing in time and a clathrate frame and that structure may include, as the main element, clusters of hydrate shells of protein macromolecules.

6.   All biological objects consist of chain-linked biological macromolecules and other components providing their particular functioning.

7.   There may be several interconnected systems in an organism, each one having a specific specter of absorption of microwave radiation (for example, conjugated biochemical reactions). In the meantime, simultaneous impact on all systems with

lower partial probability may be considerably more effective than impact on just one of them by a higher intensity radiation. In this case it would be necessary to study an effect caused by radiation for each individual frequency and, first of all, by how many times intensity of radiation must be increased in order to achieve such magnitude of the effect that would be in place in case of a multi-frequency radiation.

Based on these general premises we can suppose that the main mechanism describing influence of a microwave radiation on protein solutions is transformation of energy of that radiation into rotational energy of water molecules with subsequent transformation of a part of that excess rotational energy through the grid structure of hydrogen links to molecules of bound (hydrated) water contained in the hydrate shell of the proteins. Because of that protein hydration increases, along with the conformational state of a flexible chain-linked macromolecule.

That mechanism may be expressed schematically as a chain of consecutive processes

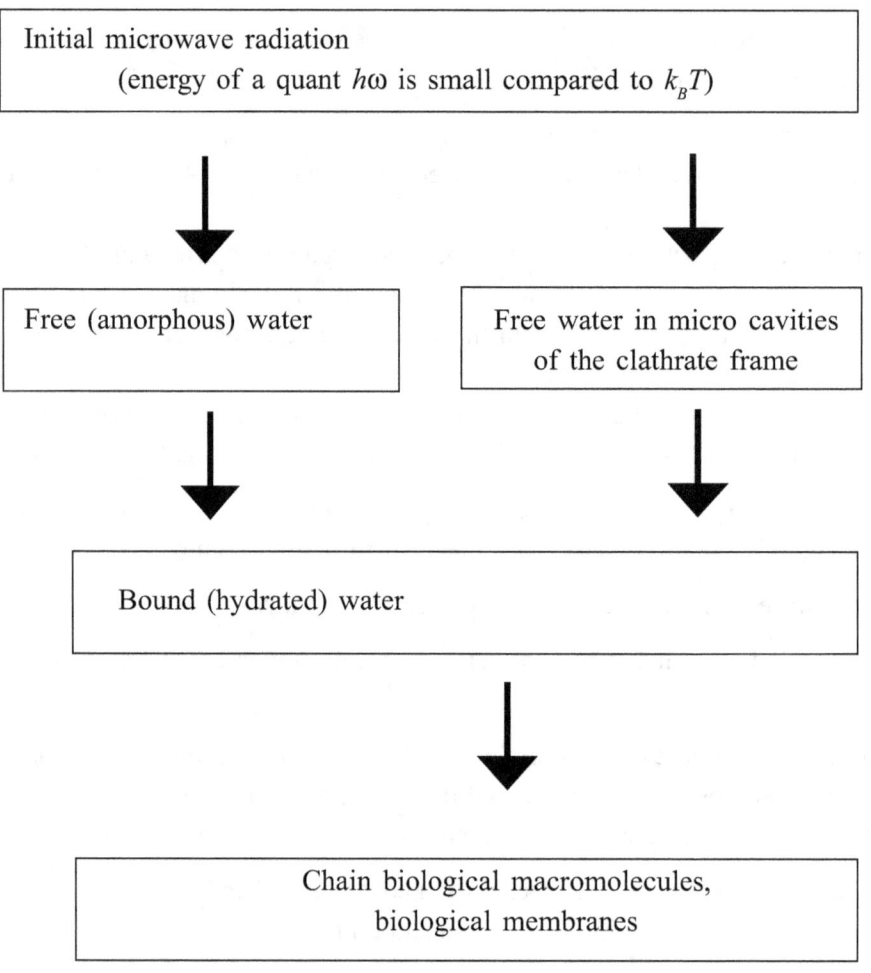

Let's consider the work pattern of that scheme.

Free (amorphous) water absorbs energy of microwave radiation on a very broad frequency range. The main mechanism of interaction is diffusion of radiation on dipole moments of separate free water molecules, invoked by disturbing of rotational degrees of freedom of these molecules.

Water molecules contained in walls of the clathrate frame are characterized by rigid links of dipole moments and, for that reason, cannot effectively absorb and diffuse microwave radiation. In contrast to that, water molecules contained in the volume of micro cavities of the clathrate frame can very efficiently interact with microwave radiation. For them, efficiency of the process of interaction is much higher than for molecules contained in amorphous water.

During diffusion elastic vibrations are being agitated, which influence molecules of bound (hydrated) water directly adjacent to biological macromolecules and effectively forming a single rigidly linked compound with them. Using different physical methods (including spectroscopy of circular dichroism, small angle diffusion of Roentgen rays, Roentgen structural analysis, *IR*-spectroscopy) has demonstrated that DNA structure in water solutions (the so-called *D*-shape) is practically indistinguishable from DNA structure in condensed state with light moistening (*B*-shape), which corresponds to relative humidity of no less than 75%. It shows that only closely located water molecules have an effect on DNA structure.

The issue of influence of water environment on biological macromolecules has recently gone through major changes. It was believed earlier that water is in fact a kind of a background (inert matrix), whose only function was, that such system provided conditions for relocation of chemical elements and biochemical components toward active surface of biological macromolecules. However, discovery of numerous factors, which required direct account of interaction of macromolecules with its hydrate-ionic environment for its explanation, for example DNA polymorphism, has mandated a radical reevaluation of such approach. There are actual facts demonstrating an active role of water in metabolism processes.

For instance, replacement of light water H_2O to heavy water D_2O in any biochemical experiment in most cases leads to fatal results for vital processes. In particular, microbiological cultures, which grow well in a nutrient medium based on H_2O, are practically unable to grow in a similar nutrient medium based on D_2O. A slight variation in density, viscosity, ionic content of these two liquids cannot explain that effect. Actually, many experiments, in which using buffer environments it was possible to correct these variations, didn't produce a corresponding leveling of growth rates of those cultures.

An obvious distinction here is that the natural spatial period of light water structure is ideally compatible with DNA spatial structure on its entire length. Meanwhile, slight difference of the spatial period of heavy water structure (owing to a different length of the O–H link, equal to 0.9572 *A* and the O–D link, equal to 0.9575 *A*) causes the situation when in every 1000–1500 periods of the spatial lattice of water the structure of water adjacent to a DNA and spatial structure of DNA itself will be mutually disordered, while at a distance equal to 3000 periods there will be one "extra" period. This circumstance causes appearance of great mechanical tensions in DNA and disrupts a normal process of protein synthesis.

An analysis of Roentgen structural data has allowed to factor out at least two main types of water molecules in a hydrate environment of double spirals. Such water may be localized on nitrous bases in a minor or main chutes of DNA in the form of "water bridges" and "water crest" and be located near a negatively charged phosphate groups. From the point of view of energetics, a water molecule may be considered bound if its energy of interaction with a hydrate-active center exceeds the average energy of interaction between water molecules in a liquid phase (about 0.46 *eV* or 43.9 *kJ*/mole).

Using the results of a Roentgen structural analysis, *IR-* and *SHR*-spectroscopy, it became possible to determine that full hydration of a double spiral in, for example, *NaDNA* in *B*-shape is possible to achieve by connecting about 20 molecules of water (or about 60 atoms of oxygen and hydrogen) per each nucleotide. Considering that there are about 25 atoms in each nucleotide, we can see that in a system (water + DNA) the former plays (quantitatively) the main role.

As we increase water content, the sample will have different mechanical properties of the water-DNA system, the elasticity factor will be lower, mobility of atomic groups in DNA and mobility of water molecules in its hydrated environment will be higher.

In some experiments there was a sharp increase of absorption of an electromagnetic radiation of the centimeter diapason by water solutions of DNA in biological culture *E.coli*. In particular, in the process of incubation of DNA with an enzymatic DNA-polymerase there was an intense additional absorption in the range of 9-12 *GHz* exceeding absorption of an equivalent amount of water by 400 times! In the case of water solutions of plasmid DNA with molecules of a certain length, the same group of researchers has discovered an intense resonance absorption in the frequency range of 2-10 *GHz* (Edwards, 1984; Davis, 1988). The experimentally observed effects were explained on the basis of the model of acoustic waves in DNA fragments of a certain length, with a special provision for suggestion about the presence of a first hydrate layer, which actively interacts with DNA.

Beside that, bound (hydrated) water is a stabilizer of the tertiary and quarternary structure of proteins and any changes in the structure of bound water itself may cause conformational changes in proteins. The process of water interaction with biomolecules in conditions of a dissolved solution presents most problems since water bound with macromolecules constitutes just a small portion of all water, not exceeding $10^{-2} - 10^{-5}\%$ of the total amount of water in a solution.

A flexible chain biological macromolecule is a big statistical system with constant average energy consisting of a large number of elements. Due to flexibility of the chain composition of those elements changes continuously. In result, a macromolecule may be characterized by a combination of conformational states, which are realized in specific conformational configurations. Rearrangement of these configurations occurs at random.

With an external influence on a chain macromolecule it may become chaotic as well as ordered depending on the character of influencing. External influence may affect the electronic or mechanical subsystems, it may change conditions of their mutual balance, alter the character of exchange of energy streams between inner subsystems due to alteration of hidden parameters.

Circulation of energy streams in statistical equilibrium between inner subsystems is the subject of synergetics. Controlling this phenomenon is a peculiar task as, for example, there may be conditions, when weak external influences can significantly change properties of chain molecules. This circulation of energy streams between inner subsystems at an external statistical equilibrium apparently plays a certain role in specific functioning of biological objects, when due to external influences these energy streams are synchronized and directed.

Based on this discourse, we can conclude that a macromolecule in a state of statistical equilibrium does not tend to become totally chaotic in any microconditional space. It's impossible to identify conditions of equilibrium with a solvent for a macromolecule without introducing new initial conditions. A macromolecule has hidden physical parameters, which make possible realization of weak informational interactions, which may cause very significant changes of its structure. As we know, such process corresponds to the effect of bifurcation.

Water plays the decisive role on all stages of these complicated transformations.

Along with specific evolution of protein macromolecules in water environment, biological membranes also play an important role in cellular metabolism. These membranes perform a number of functions, which, at a first glance, are hard to accommodate within one object. At the one hand, they need to be strong enough and perform the function of a protective barrier. At the other hand, they need to be selectively permeable, which provides for the possibility of conformational alteration of proteins contained in them. Membranes, as well as proteins, are dynamically balanced structures, while water not only participates in organization of these structures, but it also actively involved in processes occurring in them. In support of this statement we can say that the process of ionic transportation and ionic identification in membranes is impossible without water. It is evident that activated water, having different physical and chemical characteristics, can significantly modify all these processes.

Literature to Chapter 5

Davis M.E., Van Zandt L.L.// Phys. Rev. A. 1988. V.37. *P.888.*

Edwards G.S., Davis C.C.. Suffer J.D., Swicord M.L.// Phys. Rev. Lett. 1984. V. 53. *P.1284.*

Ignatiev A.M., Rukhadze A.A. // Physics of Plasma, v. 8 (1982) *p. 80*

6. IMPLEMENTATION OF MOLECULAR RESONANCE EFFECT TECHNOLOGY FOR THE ACTIVATION OF WATER BOTH FOR PHYSICAL PROCESSES AND FOR INFLUENCES ON BIOLOGICAL SYSTEMS.

There are lots of types of purified waters such as spring, distilled, colloidal, and nanoclustered waters. But no process has previously been known which can alter the molecular structure of water without any foreign substances being introduced into the water. This process is called Molecular Resonance Effect Technology (MRET) and it was patented in the USA in February 2000, US patent No. 6022479 — *Method and Device for Producing Activated Liquids and Method of Use Thereof*. This invention relies on the idea that electromagnetic radiation can affect the atomic and molecular structures of substances. This fact was proved by specific class of experiments involving Rydberg atoms - atoms with an electron in a highly extended orbital (DUNNING, Barry, F., Rydberg Atoms - Giants of the Atomic World, *Science Spectra* issue 3, pp. 34-38, 1995).

The effect an electromagnetic force has on an atom depends on the atom's electronic structure during the interaction. One could imagine that the application of the appropriate time-dependent force to an atom could alter its electronic structure in a specific way, thereby controlling its response to subsequent radiative or collisional processes. Furthermore, the specificity of certain reactions of electronic structure might be exploited to reconstruct the motion of the atomic electron cloud. The key to the manipulation of electronic structure in atoms is the generation of electromagnetic fields or radiation that will push and pull the electronic wave function in a controlled and reproducible way. (JONES, Robert, R., Modifying Atomic Architecture, *Science Spectra* Issue 22, pp. 52-59, 2000).

The water molecule has a polar triangle structure with covalent bonding of two hydrogen atoms to one oxygen atom. There is a measured 104.5° angle between these bonds. Water is one of the most polar molecules known in nature. The polarity of water underlines its chemistry and thus the chemistry of life. Polar molecules interact with one another through attraction. This weak attraction is called a hydrogen bond. In regular water polar molecules form short-range, unstable associations of different crystal shapes. Untreated water is able to form liquid crystal associations that have only 5 to 10% of the strength of covalent bonds. According to proposed hypothesis, the process of water activation induces the formation of long-range water molecular domains similar to water molecular structures found in living cells.

The water-activating device used in Molecular Resonance Effect Technology is made of a polar polymer compound with long linear molecular structure mixed with certain amounts of physiologically active organic and inorganic substances. Most polar polymers possess comparatively high values of relative permittivity (dielectric constant). This means that external electromagnetic forces can easily displace both bonding and non-bonding electrons in the molecular structure of these polymers.

While many polymers are highly flexible and form an amorphous solid upon the process of polymerization, a large number of polymers such as epoxy actually form partially crystalline structures. Epoxy is formed by mixing Bisphenol A with low-molecular weight liquid resin that contains epoxy groups. The principal reaction of epoxy groups with phenolic hydroxyl functions leads to linear polymer chains formation (nM —— Mn, where $n > 38$) (*Fig. 6.1*).

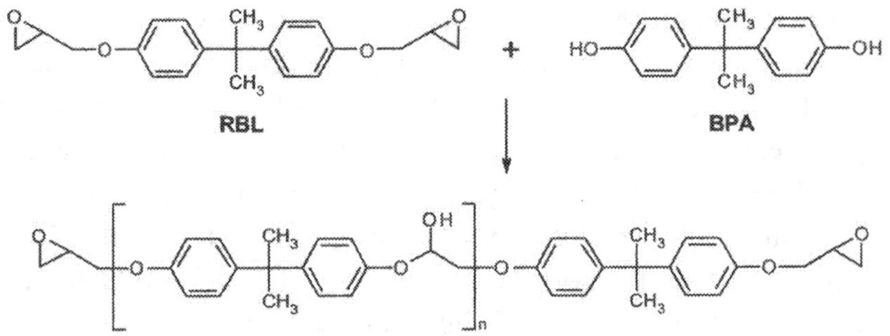

Fig. 6.1 Epoxy polymer typically contains highly polar hydroxyls and amines. Once all the amines sites have reacted with the epoxy sites a three dimensional network is achieved.

In case of epoxy polymer the kinetics to a large extent determines the final crystalline structure of the polymer. The introduction of foreign (physiologically active) agents in the parent lattice of epoxy polymer leads to effect of superimposed periodicity and as a result develops modulated crystalline structures with specific microstructure, phase transition, network topology and polarity.

A number of studies show that external electromagnetic field can affect local orientations and phase transitions in polymer crystalline systems of longitudinal chains. The longitudinal polymer crystalline system is a macromolecule of consecutively copolymerized liquid crystals and flexible polymer sequences. The external electromagnetic field can seriously modify the local orientation order of the system and affect phase transition parameters and dielectric properties of the polymer compound. A simple molecular mechanism exists since the polar parts of the molecule in epoxy are rigidly attached to the chain backbone. The orientation of the polar groups in electromagnetic field affects the backbone orientation. The extend of the local orientation of crystalline structure of epoxy introduced to electromagnetic field has been determined with an anisotropy parameter, based on the Ultrasound Critical-Angle Reflectometry.

In order to achieve the activation effect in liquid water, the polymerized epoxy with incorporated physiologically active agents is exposed to homogenous magnetic field and oscillating optical light of 600–700 *nm* wave length and 7.8 *Hz* frequency. The external electromagnetic field generates an excitation in crystalline structures of the polymer compound. The existence of orientations and phase transitions in crystalline systems of epoxy polymer introduced to external electromagnetic field leads to the origination of subsequent relaxation and strain phases in macromolecular structures that induces the phenomenon of piezoelectricity in polymer compound. Due to this phenomenon of piezoelectricity the epoxy polymer compound generates subtle electromagnetic oscillations that activate liquid substances, particularly water. There is no direct contact between polymer compound and the body of treated water. The distance between polymer compound and the surface of water is at least 1 (one) inch.

According to proposed hypothesis the process of activation modifies the configuration of molecules and hydrogen-bonding patterns in water. As a result, the water molecular organization state is changed, and the long-range water molecular domains are formed.

The mechanism that explains the effect of electromagnetic fields on water is related to the existence of defects in molecular structure of water. The stable structural changes in water were detected in experiments by the *UV* luminescence spectrophotometer. They have been attributed to different water structural defects that include specific centers of luminescence. The nuclear proton spins were considered to be a primary targets of external magnetic fields, since proton lattice of water molecules is unstable and asymmetric. The structural metastability of water was associated with microscopic orbital currents of protons in water-molecular hexagons, and deviation from the stochiometric composition of water. The effects of memory of water interacting with electromagnetic fields were supposed to originate from the oscillations of water-molecular hexagons.

Bernal-Fowler rules describe structural organization of hydrogen and oxygen atoms in the system of hydrogen bonds in water. According to these rules: a) each oxygen atom has covalent bonding with two hydrogen atoms; b) each hydrogen bond (O–O line) has one hydrogen atom. The dissociation of water molecules abuses Bernal-Fowler rule a), and leads to the existence of hydroxyl ions OH^- and hydroxon ions OH_3^+ and as a result ionic defects have place in the molecular structure of water:

$$2H_2O \leftrightarrow OH^- + OH_3^+$$

The deviation of the equilibrium of ionic defects in water can be easily reached by adding acid or alkaline in the body of water.

There are also Bierum orientation defects existing in water molecular structure that abuse Bernal-Fowler rule *b*), and lead to the existence of *D*-defect (there are two protons on line O–O), and *L*-defect (there are no protons on line O–O). These defects appear as a result of disorientations of molecules in the hydrogen bonding of water *(Fig. 6.2)*.

$$2N \leftrightarrow D + L$$

where N – hydrogen bonds without defects.
L-defect has negative electrical charge, and *D*-defect has positive electrical charge.

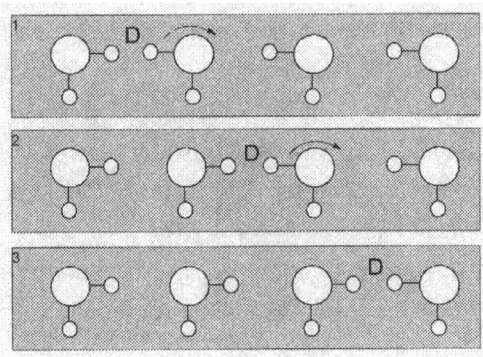

Fig. 6.2
The transference of D-defect as a result of proton
rearrangements in molecular structure of water
(Binhi, 1998).

The recombination of Bierum orientation defects has no direct correlation with ionic defects in molecular structure of water. The orientation defects in water can take place in case of deviation of stochiometric composition of water. The *pH* and *NMR* tests conducted on Activated Water show that the process of activation has tendency to develop both ionic and orientation structural defects in water.

The orientation of nuclear proton spins may influence biochemical processes in biological systems. It is a result of associations and disintegrations of mentioned above structural defects of water, since ionic structural defects are chemically active. The recombination of water defects is classified within both classical and quantum types of description. The nuclear proton spins exposed to the external resonance magnetic field are loosing their dynamic correlation that leads to recombination of water defects and to deviation of stochiometric composition of water. The recombination of water defects in magnetic field is a result of proton spin orientations that initiates the quantum transition of proton from one potential position to another potential position in the lattice of hydrogen bonding in water. In this case the potential energy of proton is characterized with 'two position' curve *(Fig. 6.3)*. (BINHI, V.N. The structural defects of liquid water in magnetic and electrical field. *Biomedical Radioelectronics,* 1998, no.2, p.7-16.)

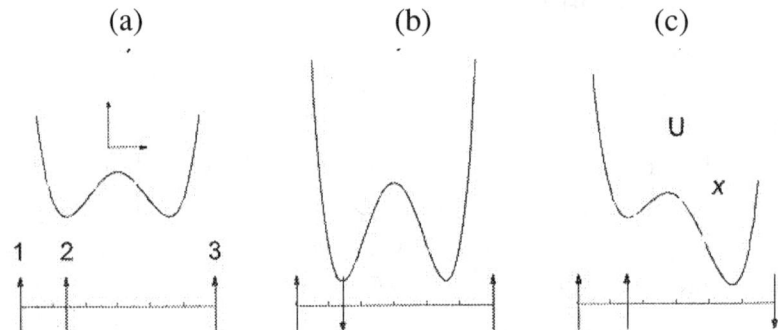

Fig 6. 3   The variation of potential positions of proton as a result of different
nuclear proton spins orientations (Binhi, 1998).

The quantum transition of protons leads to the fluctuation of the probability distribution of nuclear proton spin functions. For instance, the probability distribution of nuclear proton spin projections at the direction of magnetic field is different for static equilibrium and for magnetic resonance *(Fig 6.4)*.

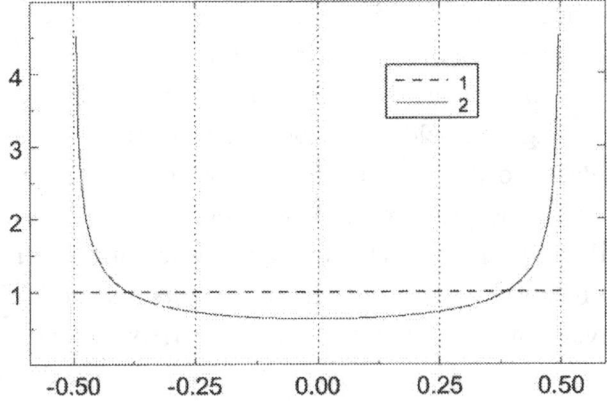

Fig 6.4   Density of the probability of proton spin projections in case of:
1-thermodynamic equilibrium; 2-magnetic resonance (Binhi, 1998)

In case of resonance magnetic field, nuclear proton spins have tendency to orient themselves along the lines of magnetic field. In this case proton spins are oriented in predominant directions for a long period of time and their precession is slowing down or even stops. This situation may lead to formation of stable structural defects in water.

The probability of the quantum transition of protons depends on the interaction of the wave functions of the neighboring protons that in its turn is a result of nuclear proton spin orientations. The oscillating magnetic field can influence the probability of proton quantum transitions. In theory, in the system exposed to non-resonance magnetic field the probability of proton quantum transitions can be considerate time-independent factor P_0. In the system exposed to resonance magnetic field the probability of proton quantum transitions is time-dependent factor $P_{(t)}$. The mean value of the intensity of proton quantum transitions in the system exposed to resonance magnetic field differs in time from the intensity of proton quantum transitions under influence of non-resonance magnetic field. The theoretical calculation reviles the following result (Binhi, 1998):

$$(P_{(t)} - P_0) : P_0 \approx 41\%$$

It shows that nuclear proton spins in the lattice of hydrogen bonds in water can be controlled by the resonance magnetic field applied to the system.

The stable water molecular clusters were discussed based on observed low-frequency spectra of the water electric conductivity in a number of experiments. The clusters were assumed to memorize an electromagnetic activation in water molecular structure. Thermal fluctuations of the kT scale are ten orders greater than the quantum of the hydrogen bonding energy. In this case, the question is important – why do these random thermal disturbances not destroy the molecular clusters of water? A suggested solution of the problem is related to the idea of coherence of the external stimulus (resonance magnetic field), against the background of incoherent thermal noise. As a result high quality molecular oscillators, such as water-molecular hexagons, may be swung up in a time-space coherent manner to a condition when the quantum of collective excitation energy will be predominant over the energy of random thermal fluctuations. .(FROHLICH,H. and KREMER,F. *Coherent Excitations in Biological Systems.* Springer-Verlag, New York, 1983.)

Thus, specially modified magnetic field can induce the formation of metastable structural composition of water that is related to the intensity of protons recombination in the lattice of hydrogen bonding in water. One of the primary magneto biological mechanism associates the effects of subtle magnetic fields with altered states of liquid water in biological systems. The structural changes in water that result from the influence of external magnetic field are further transmitted to the biological level, since water takes part in a variety of metabolic reactions.

Nuclear Magnetic Resonance, Dispersion Staining Microscopy and Laser Spectroscopy tests confirm changes in the molecular structure of the water molecules in Activated Water.

NMR tests were conducted at the laboratory of Department of Chemistry of San Diego State University by Dr. Leroy Lafferty. The experimental data were recorded on INOVA-500 spectrometer. Each sample of Activated and regular water was inserted into 10-*ml NMR* test tube, and a 5-*ml NMR* tube filled with D_2O was then inserted into the 10-*ml* test tube to provide a lock signal for the *NMR*. A 90° second pulse was used for the experiment. Acquisition time was set to 5.000 sec with one second delay, and a spectral width of 8000.0 *Hz* was employed. Line broadening was utilized and set to 0.2 *Hz*. Fourier transformation was performed on each spec following the scanning. Signal peak was observed for all samples, indicating that both Activated and regular water samples are free of detectable organics. Experimental data revealed a consistent 2.1 times increase in the width of proton peak in the line of *NMR* absorption for the samples of Activated Water in comparison with a sample of regular water from the same source *(Fig 6.5)*.

Another *NMR* test was conducted by Dr. Lin Chiang at NuMega Resonance Laboratory. Experiment was conducted in compliance with standard *NMR* methodology. The data were recorded on a 500 *MHz* Bruker *AM*–500 spectrometer. The tests showed a consistent 2.5 times increase in the width of the proton picks in the line of *NMR* absorption for all types of Activated Water *(Fig 6.6)*.

(A)

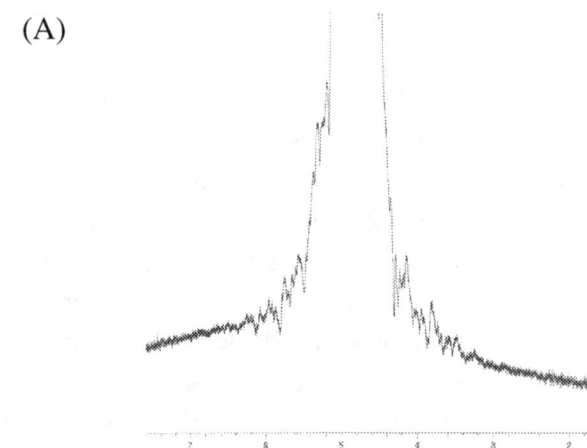

(B)

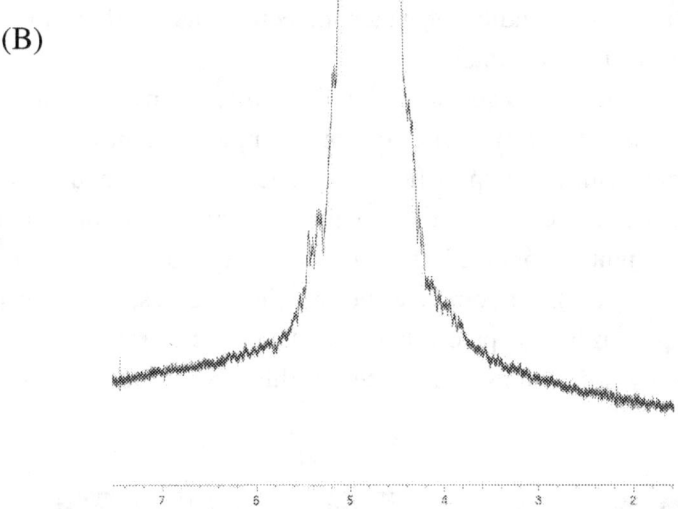

Fig 6.5 NMR test result: A- sample of Activated water, and B- sample of regular water. Observe H-1, 499.9189962 MHz for both samples. Proton dispersion in Activated water sample is increased.

(A)   (B)

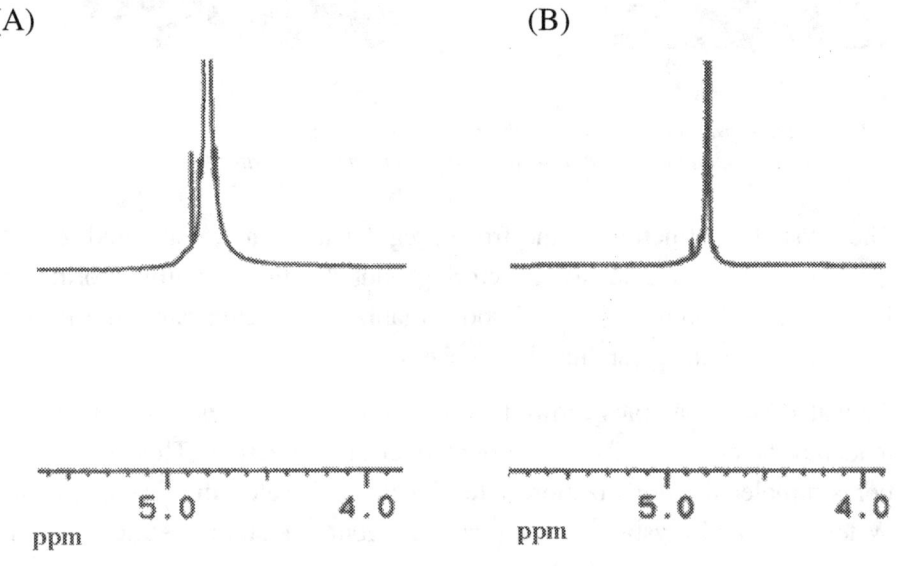

Fig 6.6 NMR test result: A – sample of Activated water; B – sample of regular water. Proton dispersion in Activated water sample is increased.

According to Nuclear Magnetic Resonance Theory there is a synonymous correlation between the form of the line of *NMR* absorption in the homogeneous magnetic field and the characteristics of molecular motions and dispersion in the liquids being tested. Therefore, the increase of proton pick width and proton dispersion could occur only as a result of proton recombination in the lattice of hydrogen bonding of water molecules that leads to the deviation of stochiomentric composition in Activated Water.

Analyst Bryan Burnett conducted Dispersion Staining Microscopy tests according to standard methodology at Meixa Tech Laboratories. The six water samples (3 control and 3 activated from the same source) were frozen with the help of liquid Nitrogen injection, and then analyzed in polarized light. Crystals diffract light depending on structure, thickness, and orientation to the incident light. When the source of light is polarized, the light emerging from the crystal is broken down into restricted wavelengths (colors). To complete the imaging process, a second polarizer, that is cross-oriented to the first polarizer, was placed between the crystal sample and the camera. Dispersion Staining Microscopy tests did show differences in this experiment *(Fig 6.7)*.

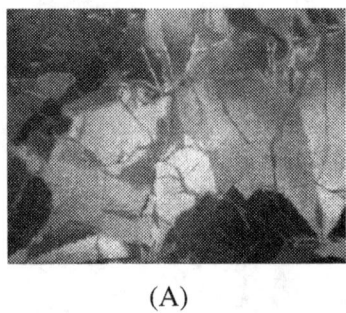
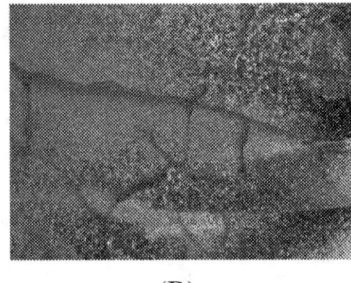

(A)   (B)

Fig 6.7   A – the crystalline structure of the frozen tap water;
B – the crystalline structure of the frozen Activated tap water.

a)   The crystalline structure of the frozen regular tap water, compared to the Activated tap water, showed extensive fracturing and chaotic crystalline formation. This is likely due to the impurities and poor organized molecular structure in the tap water that interfere with crystallization process.

b)   Activated Water samples showed well organized crystalline formations with a strong tendency of crystal axis to be oriented in one direction. This is likely a result of domain molecular organization state of Activated Water. In this experiment Activated Water generated crystals with typical hexagonal crystalline structure. This fact also illustrates that Activated Water has fewer impurities than control tap water samples.

The Laser Spectroscopy test was conducted according to the standard methodology by Professor S.G. Alexeev and Dr. A.A. Kornilova at the laboratory of Institute of Optical – Physical Measurements of Russian Academy of Sciences. The Laser Spectroscopy method is based on the measurement of the density of light diffraction in the liquid media. When the laser beam goes through liquid media its diffraction can be changed depending on the size and configuration of molecules. Based on this idea the device can be calibrated to detect any changes in molecular structure of water. Three different samples were tested:

a)   sample of triple-distilled water as a reference;

b)   sample of regular tap water as a control sample; and

c)   sample of tap water from the same source after 15 minutes of activation.

Frequency Spectrum of Intensity of Fluctuations
(autocorrelation Furies-image function)

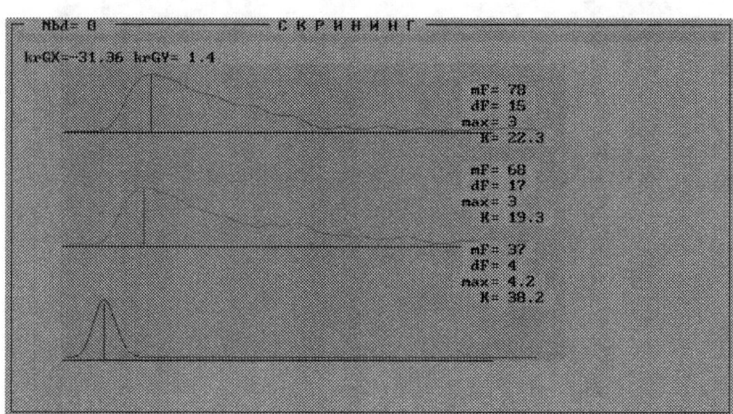

*Fig 6.8   Top chart presents Laser Spectroscopy result for regular tap water (control);
chart in the middle presents Laser Spectroscopy result for Activated tap water;
and chart at the bottom presents Laser Spectroscopy result for triple-distilled
water (reference).*

The comparison of charts for regular and Activated water reveals that activation process actually changed molecular structure of water. The shift of maximum pick of Activated water chart to the left direction (*mF* = 78 regular water; *mF* = 68 activated water) and the increase of the square of Activated water chart confirm changes of physical properties of water that could occur only as a result of structural changes in Activated water.

Another experiment that proves the ability of activation process to form structural changes in water was conducted on concrete at Geotechnical and Environmental Sciences laboratory in San Diego. This test was conducted according to the standard methodology for verification of compressive strength and total load capabilities for type II cement mixed with tap water and Activated tap water respectfully *(Fig 6.9)*. The two concrete trial batches including the fabrication of six concrete compressive strength cylinders were performed. Both batches were tested for slump and unit weight. The slump tests were performed in general accordance with ASTM C143-90. The unit weight tests were performed in general accordance with *ASTM C*138-92. Six compressive strength cylinders were cast from each batch and cured in a moist curing room in general accordance with ASTM C31-96.

COMPRESSIVE STRENGTH TEST

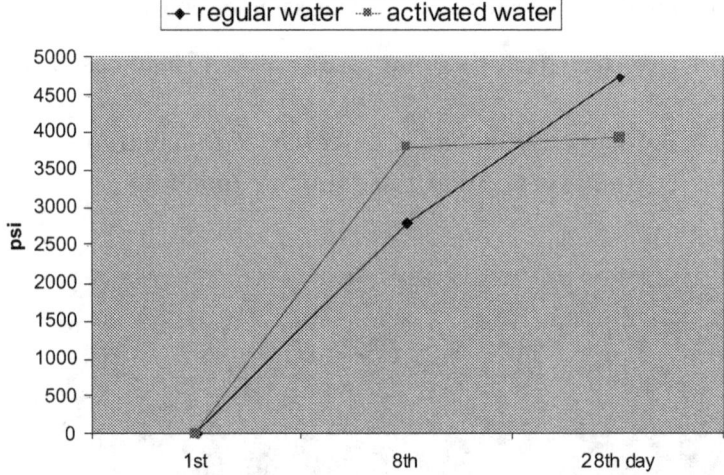

Fig 6.9   Compressive strength test reveals that Activated water provides accelerated hydration reactions.

When the cement is mixed with water, hydration reactions convert the water-cement suspension into rigid porous material that serves as the matrix for concrete. This experiment showed that Activated Water accelerates hydration reactions. The set point for Activated samples of concrete was reached 3.5 times faster then for regular samples. Taking into consideration, that no accelerators were added to Activated Water, and the ratio of water to cement for all tested samples was the same, it is reasonable to admit that acceleration of hydration reactions in activated samples is a result of specific structural changes in Activated Water.

The process of activation in compliance with MRET affects liquid substances other than water in a similar way, by changing their molecular structure. The experiment was designed to find out the effect of activation process on polyurethane material. Polyurethane is formed due to a chemical reaction between a polyol resin and a polyisocyanate hardener. When two liquid components are mixed, the hydroxyl groups (–OH) in the resin react with the isocyanate groups ($N=C=O$) in the hardener, and the three dimensional molecular structure is produced. Only one isocyanate group should react with one hydroxyl group in order to provide an ideal ratio of hardener molecules to resin molecules which will give optimum mechanical properties of the final product. In this experiment resin and hardener were activated for 45 minutes before mixing procedure and polymerization reaction. Then mechanical properties of activated polyurethane samples were compared with mechanical properties of regular polyurethane. It was found that the activation process dramatically changed mechanical properties of polyurethane. Activated polyurethane became a porous, high-surface-area material, more flexible then the regular polyurethane *(Fig 6.10)*. Relatively close result in modification of mechanical properties of polyurethane can be achieved by variation of mixing ratio of hardener and resin. The molecules of isocyanate in the hardener are capable of cross-linking with itself. As a result of such process polyurethane is more flexible when the volume of hardener is less then the volume of resin.

Activated polyurethane   **Regular polyurethane**

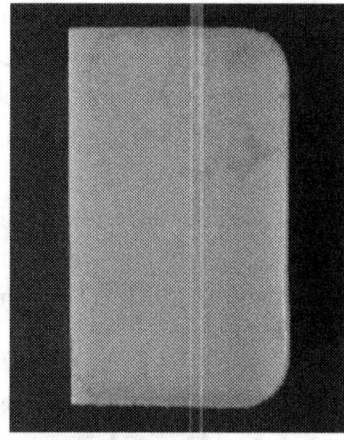

Fig 6.10 Mechanical properties of polyurethane were changed after the process of activation.

This experiment shows that the activation process affects polymer dynamics by changing associated connectivity and structure of macromolecules in polyurethane.

The benefits of Activated Water have been confirmed by extensive experimental work. Tests on Activated Water show the effect of balancing *pH* index, decrease of electrical conductivity, and significant reduction of bacterial counts. These results may be explained by the observed tendency of the activation process to create ionic and orientation structural changes in water.

Lori Motil, RM and CLS conducted these tests at CAI Environmental Laboratory. Samples of Activated Water and regular water from the same sources were analyzed utilizing EPA or other ELPA approved methodologies:

- The 15 minutes activation of alkaline water sample changed *pH* from 7.69 to 7.48, which means the 30% reduction of alkalinity (the *pH* index difference from pH_0=7.0). The 30 minutes activation showed the 62% reduction of alkalinity; *pH* changed from 7.65 to 7.25. The 30 minutes activation of acid water sample changed *pH* from 6.73 to 6.89 that mean the 60% reduction of acidity.

pH Test

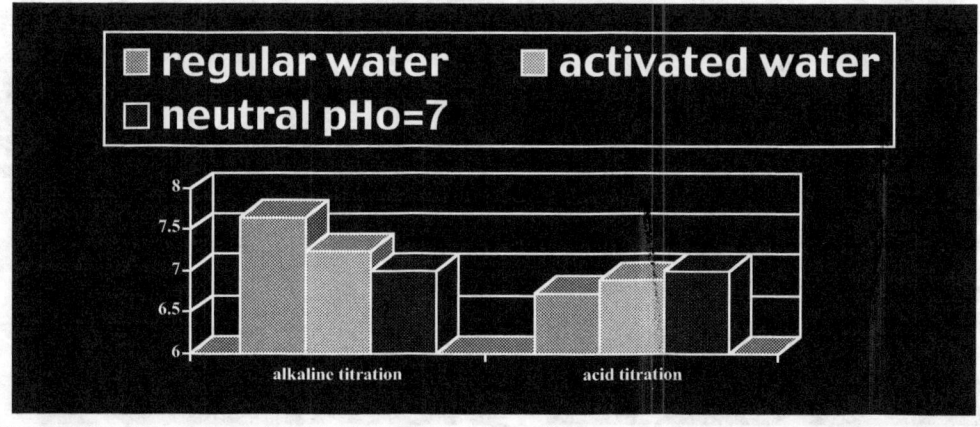

It is significant that these tests showed the tendency of activation process to balance the *pH* index to $pH_0=7.0$, reducing both acidity and alkalinity. It may occur because the free radicals of H^+ and OH^- are going through more intensive recombination with long-range water molecular structures with modified hydrogen bonds in Activated Water than with short-range molecular structures in regular water. This test clearly confirms that the activation process affects structural ionic defects in water.

- Conductivity test was conducted at Culligan Hydraulic Lab by the analyst John Van Newenhizen. Conductivity of water drecased from 720.0 MMHOS/CM to 704.0 MMHOS/CM (Method 120.1, TDS) under 27°C within 30 minutes of activation.

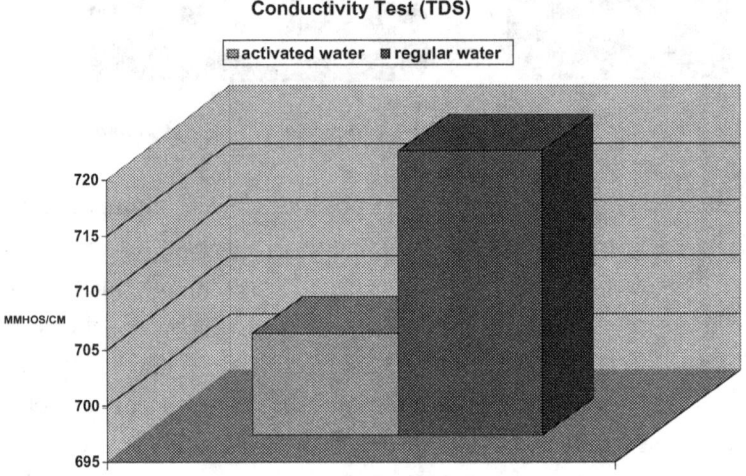

These results confirm the idea that the process of activation creates long-range molecular structures in water and leads to the deviation in stochiometric composition of water.

- The microbiology test showed the 86% decrease of total and fecal coliforms in the rainwater activated for 30 minutes. The heterotrophic plate count test showed the 44% decrease of bacterial colonies in the lake water activated for 15 minutes.

Microbiology Test

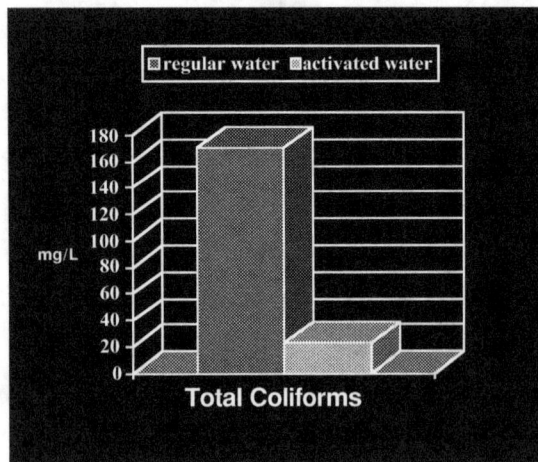

Heterotrophic Plate

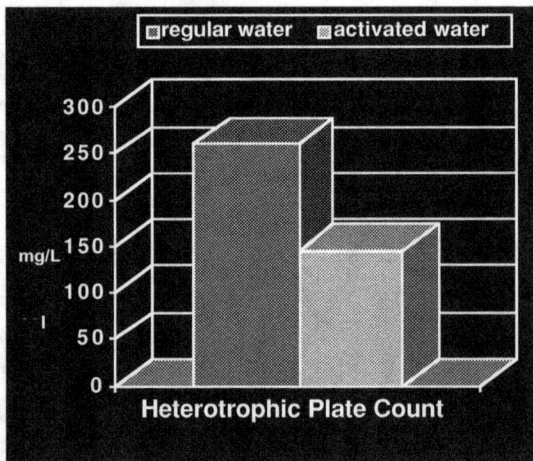

These tests prove the sterilization effect of the process of water activation on harmful microorganisms. They demonstrate that the structural changes in water resulting from the influence of external magnetic field are further transmitted to biological systems, since water takes part in a variety of metabolic reactions.

The biological effect of Activated Water on living systems was confirmed by a number of experiments in vitro and in vivo.

- Dr. John Stelle conducted in vitro tests on Lymphoma cells at the Laboratory of Engene Biotechnologies Inc. The mutated cells were incubated with Activated Water for 24 hours. Sensitivity of the tumor cells to therapeutic agents (Activated Water) was determined by an in vitro sensitivity assay. The tumor cells were separated to a single-cell suspension by passage through a wire mesh, resuspended and washed in RPMI 1640, 20% fetal calf serum media and distributed at 10^5 cells per wall in 96-well plates. Cells were incubated with Activated Water and tested at concentrations ranging from 1:10 to 1:1280 for 18 hours at 37°C, and 8% CO2.

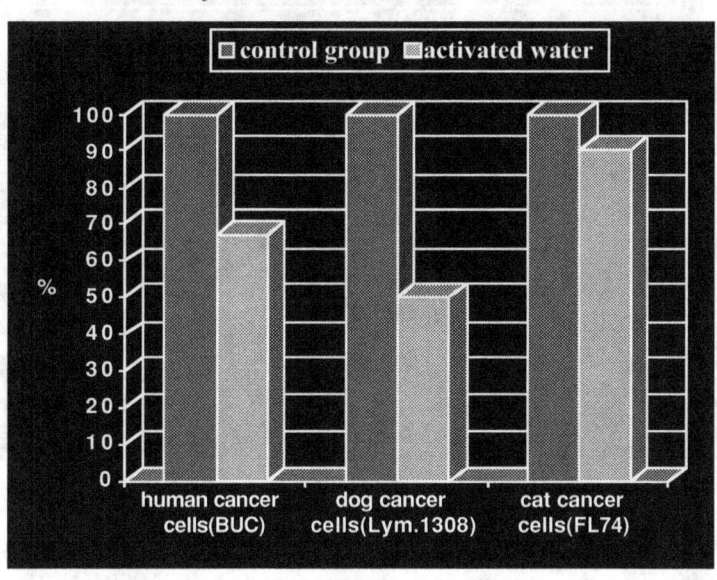

Sensitivity of Tumor Cells to Activated Water

At the end of the incubation period the imaging dye MTT (Thiazolyl Blue) was added, and the cells were incubated a further 6 hours. Cells were lysed with 1% sodium dodecyl sulphate, 5% acetic acid solution, and the absorbance was read at 700 *nm*. Percentage inhibition of cell growth caused by Activated Water was calculated relative to untreated control wells. These tests consistently showed that Activated Water suppressed the metabolism of 33% of human cancer cells (BUC), 50% of dog cancer cells (Lymphoma 1308), and 10% of cat cancer cells (FL74). The results on human and dog cancer cells were statistically valid, $p < 0.01$. The survival level of cancer cells in control groups remained 100%. Tests also showed that Activated Water is nontoxic.

- Another experiment was conducted to find out the rebounding effect of Activated water on the White Blood Cells Counts (WBCC) of patient undergoing the chemotherapy treatment. The research was based on the results of blood tests (4 chemotherapy courses) of the volunteer, a patient of Cedar-Sinai Comprehensive Cancer Center in Los Angeles with metastasized naso-pharyngeal cancer. He was taking Activated Water while going through his regular chemotherapy treatments (Taxotere chemotherapy) in September of 1998, October of 1998, March of 1999 and April of 1999 respectively.

Rebounding Effect of Activated Water on WBCC

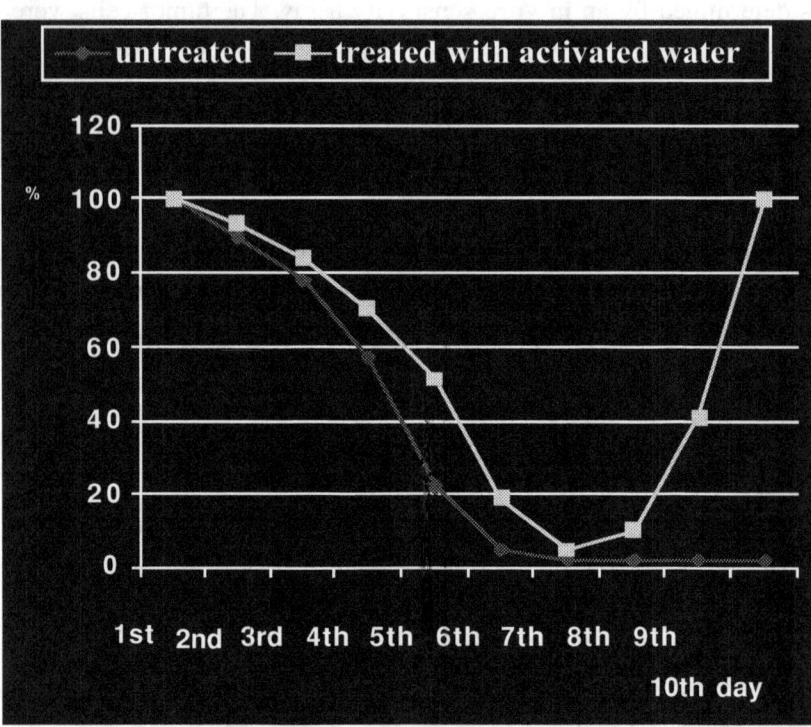

On the sixth-seventh day after chemotherapy the WBCC usually decrease from the normal range of 4.00–10.81000/U to the extremely low range of 0.1–0.21000/U (2%–3% of their pre-chemotherapy level). The rebounding period takes on the regular bases about four to six weeks. The ingestion of Activated Water prevented the decrease of WBCC to their lowest levels and helped to regain the pre-chemotherapy level in unusually short period of time. In this particular case WBCC dropped to the level of 0.5–0.61000/U (5%–9% of their pre-chemotherapy level) and rebounded to the normal level in two-three days. The rebounding effect was statistically valid, $p < 0.05$. Thus, the ingestion of Activated Water compensates one of the major side effects of chemotherapy treatment – the dramatic long-term decrease of WBCC, as well as general weakness, headache, nausea, etc.

- Live Blood Cell Analysis was conducted by Dr. Vincent Seet, Ph.D at Elixir Health

- Live Blood Cell Analysis was conducted by Dr. Vincent Seet, Ph.D at Elixir Health Ltd. headquarters in Singapore *(Fig 6.11)*. The experiment was conducted according to the standard methodology. The tested subjects were verified not to ingest Activated Water for the last 24 hours. A drop of blood sample was taken from his finger tip and placed on a specimen slide in a series of layers. After the layers dried, they were observed under the microscope. Digital camera attached to microscope took a picture of the samples. Then the subjects drank a glass of Activated Water. 20 minutes later, a drop of blood sample was taken from the finger tip one more time. Samples were observed under the microscope, and digital camera took another pictures. Both images of each tested subject were compared and analyzed. The image of the blood sample taken before the ingestion of Activated Water shows the patterns known as Rouleau formation of red blood cells. Blood cells are stacked forming worm-like patterns. The presence of massive Rouleau is a result of poor protein digestion. It usually develops such symptoms as fatigue, shortness of breath, and poor blood circulation in hands and feet because red blood cells cannot carry enough oxygen.

Live Blood Cell Analysis

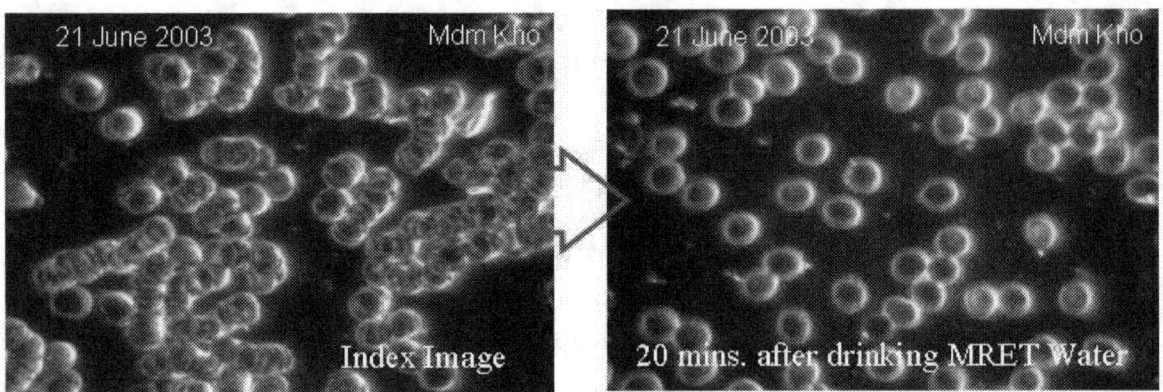

Subject 1

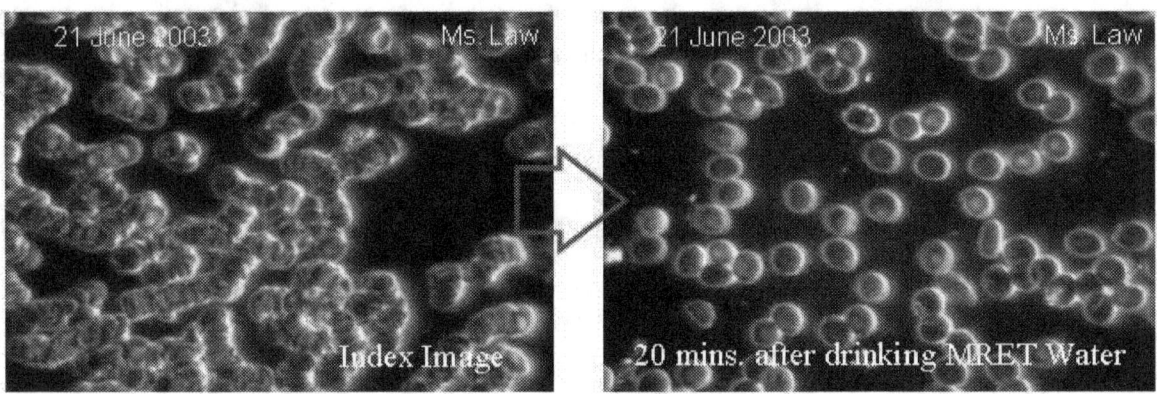

Subject 2

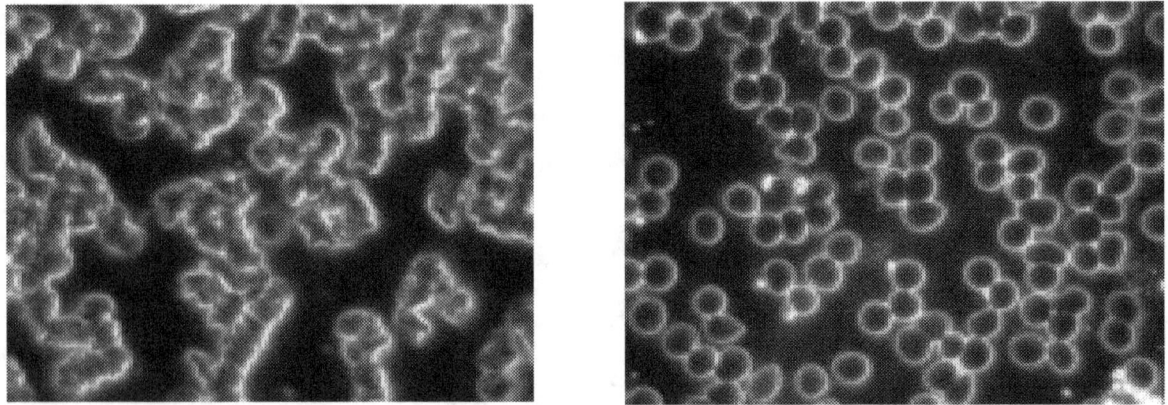

Subject 3

Fig 6.11   Left images represent blood sample before the ingestion of Activated water. Right images – 20 minutes later, after the ingestion of Activated water. It shows dramatic improvement of blood morphology within 20 minutes.

20 minutes later, after the ingestion of Activated Water the image of blood cell sample dramatically changed. The worm-like patterns were broken and the nice round individual patterns typical for healthy blood cells were formed.

- Another test was conducted on plants:
    A test was conducted over a period of 39 days to compare various plant growth using regular tap water and Activated tap water. First several packages of six different plants were selected. Seeds were planted equally in peat moss starter cups. Water was drawn from the tap and equally separated into two bottles. One bottle of water was activated for 30 minutes with the MRET activation unit. The peat moss starter pots were saturated with water with one set with regular water and the other with Activated water. It was apparent after just the first few days that the root structure of the plants irrigated with Activated water was significantly better than of the plants irrigated with regular water.

Two planter pots with normal plant potting soil were selected. Equal amounts of soil were placed in each planter. The entire peat most starter pots were transplanted into the regular planters on the fifth day taking care not to damage any roots. For the remainder of the first 32 days all plants received the same measured amounts of regular and Activated water respectively. You can see the comparisons from the pictures (regular watered plants in the left picture and Activated watered plants in the right picture) *(Fig 6.12)*.

After 32 days all watering was stopped for both sets of plants for the next seven days. After day 39 all plants were cut off at the surface and are compared in the bottom set of pictures. As can be seen the Activated watered plants grew about 50% larger during the same period *(Fig 6.13)*. Also the plants were weighed and the plants irrigated with Activated water had 50% more fiber mass *(Fig 6.14)*.

These tests are easily conducted and have been done many times in the past by others. The results are always the same. The plants irrigated with Activated water grow faster, bigger, stronger and last longer in a harsh depravation test indicating resilience.

Fig 6.12   Comparison of the plants growth after 30 days.

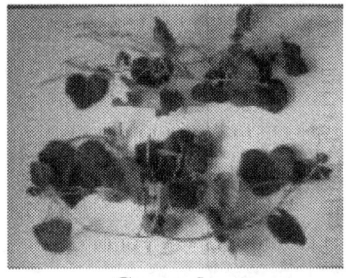  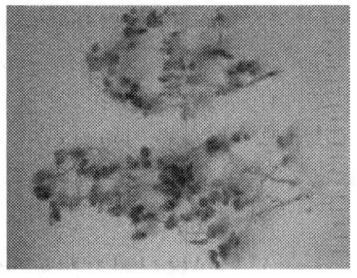

Cucumber   **Corn**   **Pears**

Fig 6.13   Graphic comparisons of plants after 32 days of watering and 7 days of water deprivation. All plants were cut off at the surface. The regular water results are on the top and the Activated water results are at the bottom.

 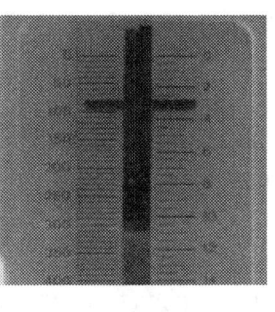    

80 grams, No Roots, Regular Water   **120 grams, No Roots, Activated Water**

Fig 6.14   The above images show the weight of the plants cut off after 39 days. The plants irrigated with Activated water were crushed down to fit into the scale. As you can see there is about a 50% increase not only is size from above but also in weight as evidenced below.

- This test was designed to compare the 15 days growth cycle of two groups of soybeans (20 beans in each group) irrigated with Activated and regular water from the same source *(Fig 6.15)*.

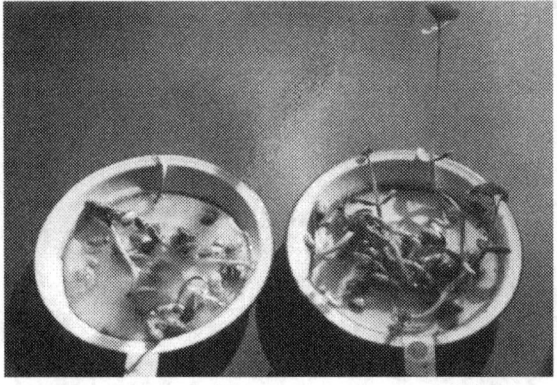

Fig 6.15   Plate with soybeans irrigated with Activated Water is marked with red dot.

By the end of this test the group of beans irrigated with Activated Water had 13 sprouts with the average length of 9". The control group of beans irrigated with regular water had only 7 sprouts with the average length of 4". This test clearly confirmed the fact of significant enhancement and acceleration of growth of plants irrigated with Activated Water. It is a result of better absorption of Activated Water because of the modified molecular structure in this water.

- Deprivation test was designed to compare the results of irrigation of plants (Tomato and Parsley) with Activated tab water (testing group) and regular tab water (control group). Plants were seeded indoors under the room temperature and irrigated for five weeks with Activated tab water and regular tab water respectively. Then the irrigation of plants in both groups was stopped for one week. The photographs of both groups of plants were made after one week of deprivation. They clearly demonstrate that plants irrigated with Activated Water developed much better resistance to deprivation stress than plants irrigated with regular water *(Fig 6.16)*.

Fig 6.16 Deprivation test on plants.

ELECTROMAGNETISM IN BIOLOGICAL SYSTEMS

It is a scientifically based fact that water plays the most important role in the vital activity of living systems, since they contain about ten thousand molecules of water per one molecule of protein. The human body, for example, is up to 75% water. It is the liquid portion of the blood and the fluid both inside and outside each cell. Water is also an important structural component of skin, cartilage and other tissues and organs. It is the medium in which the principle metabolic functions take place. It is essential for carrying nutrients, electrolytes and hormones throughout the body, and for the removal of waste from cells. Water molecules participate in cell communications and in principal metabolic functions.

The basic idea of Molecular Resonance Effect Theory is the direct transmission of prerecorded molecular activity signals to biological systems with the help of Activated Water. These messages are imprinted in water during the process of activation. Activated Water is produced from pure, clean water that has undergone a transformation of its molecular structure into organized activated state, closely resembling the cellular water. It is energized water capable of optimizing all physiological functions and carrying health-imparting messages throughout the body.

The theory of structured or domain organization and interaction of cellular water is a general consensus in contemporary scientific world. In compliance with his experimental work Dr. Ling suggested the pattern of long-range water organization in cellular systems. It includes complex multilayered water organization and long-range hydrogen bonding. Dr. Drost-Hansen, Dr. Watterson, Dr. Rorschach and Dr. Clegg confirmed that cellular water is organized into interactive domains and different liquid crystal associations traveling as a stationary wave through the medium. Based on this theory the phenomena of existence of cellular resonance was predicted in 1980 by Dr. Frohlich and confirmed later by Dr. Webb by means of laser spectroscopy.

The latest physical and biological concepts support the theory of cell communication through the transmission of electromagnetic signals (discovered by Russian scientist Alexander Gurvich in 1920's and scientifically proved by two American scientists Gilman and Rodbell. They were awarded a Noble Prize for this research in 1994).

Molecular signals are composed of low frequency waves (less than 20 kHz according to the experimental work of Dr. J. Benveniste, France) that induce cellular function and interaction. Living cells can resonate only with low frequency electromagnetic oscillations, since for millions of years of evolution they developed their normal metabolism being exposed to natural Earth geomagnetic field characterized with the spectrum of low frequency oscillations in the range of 0.5-15Hz. Living cells resonate in the range of 0.3-13Hz, and human brain function waves (Delta, Theta, Alfa) are in the range of 0.3-30Hz.

Quantum electrodynamics calls for the existence of long-range electromagnetic fields that can be transmitted by large coherent domains existing in water (DEL GIUDICE, E. and PREPARATA, E., *Journal of Biological Physics*, vol. 20, p. 105, 1994). These long-range electromagnetic fields may transmit electromagnetic signals from molecules, thus generating specific attraction between molecules with matching spectra, excluding non-resonating, unwanted random events. Therefore, specific low frequency patterns generated by defined polar polymer compounds during the activation process can be transmitted to living systems with the help of Activated Water.

Microbiology tests, Sensitivity of Tumor Cells to Activated water tests, Rebounding Effect of Activated water on WBCC test, Live Blood Cell Analysis, Soybeans Growth test, and Deprivation test on plants show that Activated Water carries and transmits the life sustaining information to cellular structures.

The explanation of this phenomenon is based in part on the possibility of the existence of resonance phenomenon between polar polymers and biopolymers such as proteins, nucleic acids, lipids, etc. The resonance effect between polar and biopolymers relies on the similarity of their long linear molecular structures and elements content. The lengths of polar polymers and biopolymers molecular structures are around 0.3-3.0 angstroms. Both biopolymers and polar polymers are composed of such elements as carbon, oxygen, nitrogen, hydrogen, phosphorus, etc.

The effect of Activated Water on molecular complexes, such as bacteria, viruses, and abnormal cells, can be explained in part by the fundamental physical phenomenon of electromagnetism, such as resonance, constructive and destructive interference. When long-range electromagnetic fields transmitted by the molecular clusters of Activated Water interact with normal cells, they create a resonance effect that enhances the biochemical reactions in these cells. When electromagnetic fields transmitted by Activated Water interact with the abnormal fields generated by the DNA of mutated cells, destructive interference is generated which suppresses the biochemical reactions in mutated cells. The same mechanism can be adopted for other molecular complexes such as viruses, bacteria, etc. The ongoing advances suggest radically new strategy to counteract the diseases caused by fault signaling in cells, such as cancer, cerebral palsy, diabetes, hepatitis C, immune system disorders, bacterial and viruses spread and reproduction.

Living organisms can use in the cell building process directly only properly molecular organized water that is similar to structured water in biosystems. Therefore, Activated Water can be easy absorbed by living systems. As a result, this water facilitates, accelerates, and enhances the process of cell building. It was proved by a number of tests presented in this article. Other forms of water, such as water with unbounded molecules or various types of water with short-range flickering hydrogen bonds, are not bioavailable to living systems. Living organisms have to metabolically restructure the water before it is bio-available to the body. Such restructuring is accomplished with great physiological 'cost'.

At the conclusion, it is possible to underline that Activated Water has outstanding physiological and physical properties. It is a result of changes in the configuration of molecules, modified hydrogen bonding and long-range water molecular structures in Activated Water. The living organisms don't have to spend a lot of energy to absorb and metabolize this water. Thus, all organs and tissues in the body receive the plentiful supply of structured water that enhances and stimulate their homeostasis.

References:

BINHI, V.N. The Structural Defects of Liquid Water in Magnetic and Electrical Field. Biomedical Radioelectronics Journal, 1998, no.2, *p.7-16.*

DEL GIUDICE, E. and PREPARATA, E. Journal of Biological Physics, 1994, vol.20, *p.105.*

DUNNING, Barry F. Rydberg Atoms-Giants of Atomic World. Science Spectra, 1995, issue 3, *p.34-38.*

FROHLICH, H. and KREMER, F. Coherent Excitations in Biological Systems. Springer-Verlag, New York, 1983.

JONES, Robert, R. Modifying Atomic Architecture. Science Spectra, 2000, issue 22, *p.52-59.*

SMIRNOV, I.V. Method and Device for Producing Activated Liquids and Methods of Use Thereof. USPTO, 2000.

www.ingramcontent.com/pod-product-compliance
Lightning Source LLC
Chambersburg PA
CBHW081125170526
45165CB00008B/2551